Recent Advances on the Use of Different Gels Type in the Food Industry

Recent Advances on the Use of Different Gels Type in the Food Industry

Guest Editors

Georgiana Gabriela Codină
Adriana Dabija

MDPI

Basel • Beijing • Wuhan • Barcelona • Belgrade • Novi Sad • Cluj • Manchester

Guest Editors
Georgiana Gabriela Codină
Faculty of Food Engineering
Ștefan cel Mare University
Suceava
Romania

Adriana Dabija
Faculty of Food Engineering
Ștefan cel Mare University
Suceava
Romania

Editorial Office
MDPI AG
Grosspeteranlage 5
4052 Basel, Switzerland

This is a reprint of the Special Issue, published open access by the journal *Gels* (ISSN 2310-2861), freely accessible at: https://www.mdpi.com/journal/gels/special_issues/DKJAUMFEN5.

For citation purposes, cite each article independently as indicated on the article page online and as indicated below:

Lastname, A.A.; Lastname, B.B. Article Title. *Journal Name* **Year**, *Volume Number*, Page Range.

ISBN 978-3-7258-5937-5 (Hbk)
ISBN 978-3-7258-5938-2 (PDF)
https://doi.org/10.3390/books978-3-7258-5938-2

Contents

About the Editors . . . vii

Preface . . . ix

Georgiana Gabriela Codină and Adriana Dabija
Food Gels: New Trends, Applications, and Challenges in the Food Industry
Reprinted from: *Gels* **2025**, *11*, 899, https://doi.org/10.3390/gels11110899 . . . 1

Hongxiao Liu, Qing Hu, Sha Yang, Lina Liu and Xuyan Dong
Temperature-Mediated Gel Texture Transformation in Starch Noodles: In Respect of Glass Transition Temperature T_g'
Reprinted from: *Gels* **2025**, *11*, 639, https://doi.org/10.3390/gels11080639 . . . 6

Chang Liu, Bing Dai, Xiaohong Luo, Hongdong Song and Xingjun Li
Polydextrose Addition Improves the Chewiness and Extended Shelf-Life of Chinese Steamed Bread Through the Formation of a Sticky, Elastic Network Structure
Reprinted from: *Gels* **2025**, *11*, 545, https://doi.org/10.3390/gels11070545 . . . 21

Luis Mieles-Gómez, Somaris E. Quintana and Luis A. García-Zapateiro
Comparison of Ultrasound- and Microwave-Assisted Extraction Techniques on Chemical, Technological, Rheological, and Microstructural Properties of Starch from Mango Kernel
Reprinted from: *Gels* **2025**, *11*, 330, https://doi.org/10.3390/gels11050330 . . . 40

Jae Young Kim and Youn Suk Lee
Self-Reactive Carbon Dioxide Absorbent with Sodium Carbonate-Based Hydrogel
Reprinted from: *Gels* **2025**, *11*, 78, https://doi.org/10.3390/gels11010078 . . . 56

Ye Youl Kim, Khawaja Muhammad Imran Bashir, Seonyeong Kang, Yongjoon Park, Jae Hak Sohn and Jae-Suk Choi
Development and Quality Enhancement of Fried Fish Cake Prototype with Transglutaminase, Trehalose, and Herbal Oil for Room Temperature Distribution
Reprinted from: *Gels* **2024**, *10*, 740, https://doi.org/10.3390/gels10110740 . . . 68

Thais C. Brito-Oliveira, Ana Clara M. Cavini, Leticia S. Ferreira, Izabel C. F. Moraes and Samantha C. Pinho
Incorporation of Locust Bean Gum and Solid Lipid Microparticles as Strategies to Improve the Properties and Stability of Calcium-Rich Soy Protein Isolate Gels
Reprinted from: *Gels* **2024**, *10*, 467, https://doi.org/10.3390/gels10070467 . . . 88

Sorina Ropciuc, Florina Dranca, Mircea Adrian Oroian, Ana Leahu, Ancuţa Elena Prisacaru, Mariana Spinei and Georgiana Gabriela Codină
Characterization of Beeswax and Rice Bran Wax Oleogels Based on Different Types of Vegetable Oils and Their Impact on Wheat Flour Dough Technological Behavior during Bun Making
Reprinted from: *Gels* **2024**, *10*, 194, https://doi.org/10.3390/gels10030194 . . . 105

Ezgi Pulatsu, Sargun Malik, Mengshi Lin, Kiruba Krisnaswamy and Bongkosh Vardhanabhuti
Preparation and Characterization of Boza Enriched with Nonfat Dry Milk and Its Impact on the Fermentation Process
Reprinted from: *Gels* **2024**, *10*, 22, https://doi.org/10.3390/gels10010022 . . . 126

About the Editors

Georgiana Gabriela Codină

Georgiana Gabriela Codină (Professor Habilitate, PhD Eng.) joined Ștefan cel Mare University, Faculty of Food Engineering (Suceava, Romania), in 2005, where she teaches different courses in the food science and technology field. She holds a PhD in industrial engineering, which she obtained in 2009, and became a PhD supervisor in food engineering in 2017, when she sustained her habilitation thesis. Dr. Codină has expertise in food rheology, food sensory analysis, bread making, beer industry, and food quality analysis through different rheological, textural and sensory methods, as well as in the design of experiments and data analysis. Her research activities focus on improving the technology of food products and the quality of different foods such as baked goods, beer, dairy products, etc. She has been in involved in more than 10 interdisciplinary research projects, has published more than 200 scientific papers, including more than 100 ISI Thomson Reuters-indexed papers, and is an author of more than 35 patents under evaluation, of which 11 have been published.

Adriana Dabija

Adriana Dabija has a PhD (2000) in industrial engineering from "Dunărea de Jos" University of Galaţi, Faculty of Food Science, Aquaculture and Fishing. Her doctoral doctoral thesis was as follows: "Biotechnology for the industrial production of the bakery yeast having a high enzyme activity". She has an engineering degree in food technology and chemistry (1986) from "Dunărea de Jos " University of Galaţi, Faculty of Food Science, as well as in aquaculture and fishing. She is also an economist, her qualifications obtained from Bucharest Academy of Economic Studies, Faculty of Management, with a specialization in business management. She is a Professor and PhD in the field of food engineering (since 2019), and a supervisor at the Doctoral School of Applied Sciences and Engineering, Ștefan cel Mare University, Suceava. She has over 39 years' experience in production, teaching and research in the field of food engineering. She has published 60 ISI Thomson Reuters-indexed papers, 178 articles in international data-based journals and proceedings, 88 papers in non-indexed scientific journals and proceedings, 4 patents, 1 Euro patent, and 31 patent proposals.

Preface

Gels, in their various forms—hydrogels, oleogels, emulsion gels and bigels—are widely used in the food industry due to their ability to improve the technological and sensory properties of products. These colloidal systems contribute significantly to the optimization of the texture, stability and nutritional value of foods, thus becoming essential components of many products appreciated by consumers worldwide.

Gel formation involves physicochemical processes such as heating, cooling or ion-induced crosslinking, each of which causes structural changes in the macromolecules that form the gel network. Through these processes, the gel-forming ingredients—usually proteins, polysaccharides or structured lipids—interact to create a three-dimensional network capable of retaining the liquid phase within it. The result is a material with controllable mechanical and sensory properties, the consistency of which can be adjusted by varying the concentration of the ingredients, the pH or the processing temperature. Thus, manufacturers can obtain gels with textures and rheological behaviors adapted to the specific needs of each food product.

The versatility of these structures gives them a strategic role not only in improving food quality, but also in developing functional or sustainable products. For example, oleogels can substitute saturated fats in processed foods, contributing to healthier formulations, and hydrogels can serve as efficient vehicles for the controlled release of bioactive compounds. In addition, bigels and emulgels combine the advantages of different types of phases, offering innovative possibilities for texturizing and stabilizing complex foods.

This Special Issue brings to the forefront the latest research results in this dynamic field. The included contributions explore multiple facets of gel science—from compositional characterization and microstructural analysis to functional, nutritional, sensory and rheological properties. Each paper reflects the efforts of the scientific community to harness the potential of gels not only as functional materials, but also as tools for transforming the paradigms of food production and consumption. By bringing together these complementary perspectives, the volume offers a broad overview of recent advances and future challenges. In a global context where sustainability and innovation are increasingly converging, gels are asserting themselves as true "soft architectures" of the foods of the future—at the intersection of materials science, nutrition and food technology.

Georgiana Gabriela Codină and Adriana Dabija
Guest Editors

gels

MDPI

Editorial

Food Gels: New Trends, Applications, and Challenges in the Food Industry

Georgiana Gabriela Codină * and Adriana Dabija

Faculty of Food Engineering, Stefan cel Mare University, 720229 Suceava, Romania; adriana.dabija@fia.usv.ro
* Correspondence: codina@fia.usv.ro; Tel.: +40-7-4546-0727

1. Introduction

Different types of gels may be used in food industry to improve the quality of food products from a technological, nutritional, and sensory point of view. These can be obtained using natural or synthetic gelling agents, such as gelatin, agar-agar, pectin, carrageenan, gums (xanthan, guar, gellan), and different types of modified starches [1,2]. Each type of gel has specific physicochemical characteristics, which benefit their different food purposes. The ability of these gelling agents to form three-dimensional networks enables the formation of gel matrices capable of retaining water within the food structure, reducing moisture loss during processing, storage, or thermal heating [3]. Additionally, gelling agents conduct an increase in the viscosity of the food products, leading to an improvement of their texture and sensory characteristics [4]. An important advantage of gels is their ability to prevent phase separation, a common issue in emulsified or dispersed food systems such as sauces, creams, and dressings, e.g., [5]. By stabilizing aqueous and fatty phases, gels help maintain product homogeneity during storage, thus preventing quality defects such as sedimentation, syneresis, or coalescence. These functional properties are particularly important in the formulation of products such as yogurts, sauces, gelled desserts, processed meats, and bakery products, where consistency, homogeneity, and shelf-life period are key factors for consumers' acceptability and the product's quality preservation throughout their supply chain [6]. Usually used for their rheological properties, such as for thickening, gelling, and stabilizing, gelling agents are now used more and more for the development of healthier and functional food formulations with specific purposes [7]. Gels based on natural hydrocolloids contribute to the fiber intake of the human body and to the controlled release of bioactive compounds. Pectin gels improve soluble fiber content from food products and can reduce the rapid absorption of sugars, with beneficial effects on carbohydrate metabolism in the human body [8]. Alginates obtained from algae can be used to encapsulate vitamins and essential fatty acids, protecting them from oxidative degradation [9]. Protein gels, such as those obtained from whey, gelatin, or vegetable proteins, contribute to an increase in protein content and may also maintain the textures of food products [10]. Starch-based gels, including resistant starches support digestive health and enhance satiety [11]. Also, gelling agents obtained from different mixes form proteins and polysaccharides may be used as hybrid gels with mechanical stability and the ability to encapsulate and release nutrients in a controlled manner [12]. Recent trends in the use of different gels in the food industry indicate the development of gelling agents such as hydrogels, oleogels, emulsion gels, or bigels for healthier food products [1,2,13]. Hydrogels are often used in deliberate fat reduction and nutrient control due to their ability to encapsulate bioactive compounds [14]. Oleogels are currently used in fat replacers, offering a healthier alternative by replacing saturated fats in food products while maintaining their sensory characteristics [15]. Emulsion gels are especially used in the development of reduced-fat food products, enhancing both stability and texture due their combination of emulsion and gel properties, stabilizing oil droplets within a gel matrix [16]. Bigels integrate emulsion and gel networks in a single gel and have the potential to encapsulate different bioactive compounds in their matrix, which may lead to the development of different functional foods [17]. Recent uses of gelling agents based on plants such as seaweed and legumes are a response to consumers' demand for various plant ingredients that improve the nutritional value of food products [10]. New trends indicate the possibility of using different gels in the 3D printing of good quality food, from sensory and nutritional points of view [18]. The future development of different gel types will lead to the creation of healthier and functional food products. However, the high cost of gelling agents such as hybrid gels or oleogels may limit commercial applications. Furthermore, the increasing demand for clean-label ingredients can represent real challenges for obtaining the desired gelling agents. Also, the stability of gels and encapsulated bioactive compounds may limit their functional use. Addressing these challenges may be the key to developing gel types that enable us to obtain healthier and functional food products.

This Special Issue, "Recent Advances on the Use of Different Gel Types in the Food Industry," provides a comprehensive overview of the current progress and emerging applications of diverse gel types in food science and technology. It highlights how various gel formulations contribute to improving food quality through technological processes and sensory and nutritional characteristics.

2. An Overview of Published Articles

This Special Issue, "Recent Advances on the Use of Different Gel Types in the Food Industry," brings together eight research articles highlighting recent advancements in gel-based systems for food applications. These researchers explored new possibilities of food quality improvement, new gel types, and their possible uses in food industry, as well as different technological changes that may occur in food processes due to the behavior of different gels.

The paper "Temperature-mediated gel texture transformation in starch noodles: In respect of glass transition temperature T_g'" by Hongxiao Liu et al. analyzes the impact of freezing temperature on the texture formation, mechanical properties, and microstructure of potato starch noodles (PSNs), high-concentration starch gels formed through gelatinization and retrogradation. Freezing below the glass transition temperature (Tg') demonstrates significant changes in the mechanical and microstructural characteristics of PSNs, which leads to more disrupted structures, while temperatures near the Tg' enhance hardness, cohesiveness, and springiness due to reduced ice crystal formation and increased starch chain mobility.

The paper "Polydextrose addition improves the chewiness and extended shelf-life of Chinese steamed bread through the formation of a sticky, elastic network structure" by Chang Liu et al. reports the effects of 3–10% polydextrose addition on steamed bread. Polydextrose addition significantly modified the dough structure and its rheological characteristics and improved bread quality, especially in terms of its dietary fiber content, chewiness, and shelf life.

The paper "Comparison of ultrasound- and microwave-assisted extraction techniques on chemical, technological, rheological, and microstructural properties of starch from mango kernel" by Luis Mieles-Gómez et al. investigates different technologies (conventional extraction, ultrasound-assisted extraction, and microwave-assisted extraction) for starch extraction in terms of the technological, chemical, pasting, gelling, and microstructural properties of starch from mango kernel. Ultrasound and microwave-assisted extraction led to a high extraction yield and new starch characteristics, which may increase its application in the food industry.

The paper "Self-reactive carbon dioxide absorbent with sodium carbonate-based hydrogel" by Jae Young Kim and Youn Suk Lee presents a new hydrogel with CO_2 absorption functionality developed through the addition of polyacrylic acid sodium salt (PAAS) to a sodium carbonate solution. The use of PAAS improves conventional sodium carbonate-based CO_2 absorbents, which require water for activation. Since the hydrogel contains water, it can react with CO_2, and the inclusion of PAAS will improve CO_2 absorption capacity.

The paper "Development and quality enhancement of fried fish cake prototype with transglutaminase, trehalose, and herbal oil for room temperature distribution" by Ye Youl Kim et al. presents a new prototype of fried fish cake with good quality and an extended shelf-life. Response surface regression analysis established optimal processing conditions for the prototypes, using whiteness, DPPH radical scavenging activity, and gel strength as dependent variables. Incorporating trehalose, transglutaminase, and herbal oils and using optimum processing conditions allow surimi-based products to be stored at room temperature, increasing their profitability.

The paper "Incorporation of locust bean gum and solid lipid microparticles as strategies to improve the properties and stability of calcium-rich soy protein isolate gels" by Thais C. Brito-Oliveira et al. investigates the addition of different levels of locust bean gum and solid lipid microparticles (SLMs) on the properties of calcium-rich soy protein isolate gel, such as in terms of its visual aspect, water-holding capacity (WHC), rheological properties, and microstructural organization. According to the data obtained, these additions indicate the possibility of improving the quality of calcium-rich soy protein isolate gels in order to be used as a new product prototype in food industry.

The paper "Characterization of beeswax and rice bran wax oleogels based on different types of vegetable oils and their impact on wheat flour dough technological behavior during bun making" by Sorina Ropciuc et al. analyzes different types of oleogels obtained from beeswax and rice bran wax and five types of oils. The study concluded that all oleogel types showed good quality characteristics for their use in the food industry and may improve the quality of bakery products since the rheological data indicate good mixing, pasting, and fermentation behavior of dough samples with the addition of oleogels in their recipe.

The paper "Preparation and characterization of boza enriched with nonfat dry milk and its impact on the fermentation process" by Ezgi Pulatsu et al. studies the impact of adding nonfat dry milk (NFDM) on boza technological process and quality. According to the physicochemical data obtained, the sample with 10% NFDM was reported as the best sample.

3. Conclusions

The research presented in this Special Issue highlights the possibility of developing new food products, such as boza-based ones, with the addition of different amounts of nonfat dry milk addition, bakery products using different oleogels types or polydextrose, or fish products combining sugars, enzymes, and herbal oils. It investigates the possibility of obtaining new gel systems in food industry that may be improved in terms of their technological process or recipe. Temperature was shown to have a significant impact on gel structure and the texture of starch-based products, while some extraction techniques, such as ultrasound, improved starch quality and processing efficiency. Meanwhile, sodium carbonate-based hydrogels emerged as promising materials for sustainable CO_2 absorption. These studies reflect the new trends in developing technologically optimized gel-based systems to obtain healthier food products with good quality characteristics.

Author Contributions: Conceptualization, G.G.C. and A.D.; writing—original draft preparation, G.G.C. and A.D.; review and editing, G.G.C. All authors have read and agreed to the published version of the manuscript.

Funding: This work was supported by a grant from the Ministry of Research, Innovation and Digitization, CNCS—UEFISCDI, project number PN-IV-PCB-RO-MD-2024-0173, within PNCDI IV.

Acknowledgments: The editors gratefully acknowledge the authors for their significant contributions to this Special Issue and their valuable research. We also thank the *Gels* Editorial Office for their support during the review and publication process.

Conflicts of Interest: The authors declare no conflicts of interest.

List of Contributions:

1. Liu, H.; Hu, Q.; Yang, S.; Liu, L.; Dong, X. Temperature-Mediated Gel Texture Transformation in Starch Noodles: In Respect of Glass Transition Temperature Tg'. *Gels* **2025**, *11*, 639. https://doi.org/10.3390/gels11080639.
2. Liu, C.; Dai, B.; Luo, X.; Song, H.; Li, X. Polydextrose Addition Improves the Chewiness and Extended Shelf-Life of Chinese Steamed Bread Through the Formation of a Sticky, Elastic Network Structure. *Gels* **2025**, *11*, 545. https://doi.org/10.3390/gels11070545.
3. Mieles-Gómez, L.; Quintana, S.E.; García-Zapateiro, L.A. Comparison of Ultrasound- and Microwave-Assisted Extraction Techniques on Chemical, Technological, Rheological, and Microstructural Properties of Starch from Mango Kernel. *Gels* **2025**, *11*, 330. https://doi.org/10.3390/gels11050330.
4. Kim, J.Y.; Lee, Y.S. Self-Reactive Carbon Dioxide Absorbent with Sodium Carbonate-Based Hydrogel. *Gels* **2025**, *11*, 78. https://doi.org/10.3390/gels11010078.
5. Kim, Y.Y.; Bashir, K.M.I.; Kang, S.; Park, Y.; Sohn, J.H.; Choi, J.-S. Development and Quality Enhancement of Fried Fish Cake Prototype with Transglutaminase, Trehalose, and Herbal Oil for Room Temperature Distribution. *Gels* **2024**, *10*, 740. https://doi.org/10.3390/gels10110740.
6. Brito-Oliveira, T.C.; Cavini, A.C.M.; Ferreira, L.S.; Moraes, I.C.F.; Pinho, S.C. Incorporation of Locust Bean Gum and Solid Lipid Microparticles as Strategies to Improve the Properties and Stability of Calcium-Rich Soy Protein Isolate Gels. *Gels* **2024**, *10*, 467. https://doi.org/10.3390/gels10070467.
7. Ropciuc, S.; Dranca, F.; Oroian, M.A.; Leahu, A.; Prisacaru, A.E.; Spinei, M.; Codină, G.G. Characterization of Beeswax and Rice Bran Wax Oleogels Based on Different Types of Vegetable Oils and Their Impact on Wheat Flour Dough Technological Behavior during Bun Making. *Gels* **2024**, *10*, 194. https://doi.org/10.3390/gels10030194.
8. Pulatsu, E.; Malik, S.; Lin, M.; Krishnaswamy, K.; Vardhanabhuti, B. Preparation and Characterization of Boza Enriched with Nonfat Dry Milk and Its Impact on the Fermentation Process. *Gels* **2024**, *10*, 22. https://doi.org/10.3390/gels10010022.

References

1. Machado, M.; Costa, E.M.A.d.; Silva, S. Soft Gels in Food Systems: Recent Advances, Applications, and Technological Innovations. *Gels* **2025**, *11*, 667. [CrossRef] [PubMed]
2. Liu, B.; Yang, H.; Zhu, C.; Xiao, J.; Cao, H.; Simal-Gandara, J.; Li, Y.; Fan, D.; Deng, J. A comprehensive review of food gels: Formation mechanisms, functions, applications, and challenges. *Crit. Rev. Food Sci. Nutr.* **2024**, *64*, 760–782. [CrossRef] [PubMed]
3. Sanpedro-Díaz, M.; García-Hernández, A.B.; Gómez-Gómez, A.L.; Salgado-Cruz, J.; Ramos-Monroy, O.A.; Oliver-Espinoza, R.; Rivera-Vargas, G.A.; Salgado-Cruz, M.d.l.P. Trends and Future Perspectives of Polysaccharide-Based Bigels from Seeds, Vegetable Oils, and Waxes: A Bibliometric Review. *Gels* **2025**, *11*, 413. [CrossRef] [PubMed]
4. Liu, L.; Gao, Z.; Chen, G.; Yao, J.; Zhang, X.; Qiu, X.; Liu, L. A comprehensive review: Impact of oleogel application on food texture and sensory properties. *Food Sci. Nutr.* **2024**, *12*, 3849–3862. [CrossRef] [PubMed]
5. Bai, L.; Huan, S.; Rojas, O.J.; McClements, D.J. Recent innovations in emulsion science and technology for food applications. *J. Agric. Food Chem.* **2021**, *69*, 8944–8963. [CrossRef] [PubMed]
6. Nazir, A.; Asghar, A.; Maan, A.A. Food gels: Gelling process and new applications. In *Advances in Food Rheology and Its Applications*; Ahmed, J., Ed.; Woodhead Publishing: Sawston, UK, 2017; pp. 335–353.

7. Banerjee, S.; Bhattacharya, S. Food Gels: Gelling Process and New Applications. *Crit. Rev. Food Sci. Nutr.* **2012**, *52*, 334–346. [CrossRef] [PubMed]
8. Yemenicioğlu, A.; Farris, S.; Turkyilmaz, M.; Gulec, S. A review of current and future food applications of natural hydrocolloids. *Int. J. Food Sci. Technol.* **2020**, *55*, 1389–1406. [CrossRef]
9. Ruiz, J.C.R.; Vazquez, E.D.L.L.O.; Campos, M.R.S. Encapsulation of vegetable oils as source of omega-3 fatty acids for enriched functional foods. *Crit. Rev. Food Sci. Nutr.* **2017**, *57*, 1423–1434. [CrossRef] [PubMed]
10. Munialo, C.D.; Euston, S.R.; de Jongh, H.H.J. Protein Gels. In *Proteins in Food Processing: Second Edition*; Elsevier Inc.: Amsterdam, The Netherlands, 2018; pp. 501–521, ISBN 9780081007297.
11. Shao, G.Q.; Zhang, H.; Xu, D.; Wu, F.F.; Jin, Y.M.; Yang, N.; Yu, K.J.; Xu, X.M. Insights into Starch-Based Gels: Selection, Fabrication, and Application. *Int. J. Biol. Macromol.* **2024**, *258*, 128864. [CrossRef] [PubMed]
12. Devi, N.; Sarmah, M.; Khatun, B.; Maji, T.K. Encapsulation of Active Ingredients in Polysaccharide–Protein Complex Coacervates. *Adv. Colloid Interface Sci.* **2017**, *239*, 136–145. [CrossRef] [PubMed]
13. Siddiqui, S.A.; Alvi, T.; Biswas, A.; Shityakov, S.; Gusinskaia, T.; Lavrentev, F.; Dutta, K.; Khan, M.K.I.; Stephen, J.; Radhakrishan, M. Food Gels: Principles, Interaction Mechanisms and Its Microstructure. *Crit. Rev. Food Sci. Nutr.* **2022**, *63*, 12530–12551. [CrossRef] [PubMed]
14. Cheng, W.; Wu, X.; Zhang, Y.; Wu, D.; Meng, L.; Chen, Y.; Tang, X. Recent Applications of Hydrogels in Food Safety Sensing: Role of Hydrogels. *Trends Food Sci. Technol.* **2022**, *129*, 244–257. [CrossRef]
15. Abe, A.A.; Aiello, I.; Oliviero Rossi, C.; Caputo, P. Oleogels: Uses, Applications, and Potential in the Food Industry. *Gels* **2025**, *11*, 563. [CrossRef] [PubMed]
16. Dai, W.; Xiong, H.; Sun, W.; Zhou, C.; Tan, C.P.; Li, B.; Xu, Y.J. Emulsion Gels in Food Systems: Advances, Oxidative Stability Challenges, and Mitigation. *Food Rev. Int.* **2025**, 1–22. [CrossRef]
17. Srivastava, S.; Singh, A. Bigels: A review on formulation, characterization, and application in the food industry. *Trends Food Sci. Technol.* **2025**, *165*, 105285. [CrossRef]
18. Sharma, R.; Chandra Nath, P.; Kumar Hazarika, T.; Ojha, A.; Kumar Nayak, P.; Sridhar, K. Recent Advances in 3D Printing Properties of Natural Food Gels: Application of Innovative Food Additives. *Food Chem.* **2024**, *432*, 137196. [CrossRef]

gels

MDPI

Article

Temperature-Mediated Gel Texture Transformation in Starch Noodles: In Respect of Glass Transition Temperature T_g'

Hongxiao Liu [1,2], Qing Hu [1,2], Sha Yang [2,*], Lina Liu [1,2] and Xuyan Dong [1,*]

1 College of Food Science and Engineering, Qingdao Agricultural University, Qingdao 266109, China; liuhongxiao0621@163.com (H.L.); hqingu@163.com (Q.H.); liuln1023@163.com (L.L.)
2 Key Laboratory of Novel Food Resources Processing, Ministry of Agriculture and Rural Affairs, Key Laboratory of Agro-Products Processing Technology of Shandong Province, Institute of Food & Nutrition Science and Technology, Shandong Academy of Agricultural Sciences, 23788 Gongye North Road, Jinan 250100, China
* Correspondence: shayangemail@163.com (S.Y.); dongxuyan@qau.edu.cn (X.D.)

Abstract: Potato starch noodles (PSN), a characteristic gluten-free Asian food, are essentially high-concentration starch gels (about 35% starch) formed through gelatinization and retrogradation. This study systematically investigates freezing temperature effects, particularly across the glass transition temperature, on PSN texture and microstructure. We found that fresh PSN have a freezing point of −1 °C, supercooling temperature of −4.5 °C, and a T_g' value of −3.1 °C. Freezing significantly reduced the adhesiveness of PSN and increased the hardness. During the 48 h freezing process, noodles frozen at −3 °C, the closest to T_g', exhibited the highest hardness (14,065.77 g), springiness (0.98), cohesiveness (0.93), chewiness (11,971.06), and resilience (0.84), and the least adhesiveness. PSN frozen within the range near T_g' (−3 °C) showed superior texture, continuous solid cross-section, and dense surface, attributed to the reverse transformation of starch, high mobility of starch chains, and smaller ice crystals. PSN frozen at −3 °C for 24 h displayed the most compact and desirable texture compared to the other samples. These findings deepen the understanding of the role of glass transition temperature in the texture formation of starch gel during freezing and provide valuable insights for optimizing the frozen processing of starch gel-based food.

Keywords: starch noodles; texture; enzymatic hydrolysis; scanning electron microscopy

1. Introduction

Starch noodles are made from pure plant starch (such as potato, sweet potato, mung bean, etc.) as raw materials and are a major category of Asian noodles [1]. Unlike wheat flour-based noodles, the unique raw material selection of starch noodles makes it ideal for a gluten-free diet. Compared to sweet potato starch noodles, those made from potato starch generally have a smoother surface, a more uniform and shiny appearance, a more delicate and smooth taste, and lower breakage and paste rates [2]. The production process begins with a pregelatinized starch binder paste, followed by six steps including forming the starch dough with mixed starch powder and water, extruding the dough into fine noodles, cooking, freezing, thawing, and packaging [3]. Among these, freezing plays a critical role in determining the final texture. It transforms the noodles from a sticky paste into a non-sticky, springy, and tough retrograded gel, effectively reducing adhesiveness between strands and forming their characteristic edible quality [4].

However, significant controversy exists regarding optimal freezing temperatures (reported from −40 °C to 0 °C) and durations. Yang et al. [5] prepared potato starch

https://doi.org/10.3390/gels11080639

noodles (PSN) at 0 °C for 36 h and obtained a desirable texture. Fan et al. [6] reported that starch noodles frozen at −2 °C to −8 °C for 10 h, followed by −18 °C for 6 h, had the best texture; longer freezing times reduced cooking and shear resistance. They proposed that water extruded as ice during freezing forms an "ice coat" around the noodles, greatly reducing adhesiveness and improving toughness. Sun et al. [7] found that the freezing temperature of −40 °C would destroy the internal structure of starch gel and reduce the gel strength of starch noodles. Thus, it could be concluded that starch noodles frozen for 6 h achieved higher tensile strength than those frozen for 3 or 24 h. Nguyen et al. [8] evaluated the edible quality and resistant starch content of starch noodles affected by different freezing temperatures (−5, −10, −15, and −20 °C), suggesting that freezing at −10 °C for 18 h is recommended to produce mung bean starch noodles with high resistant starch content and good quality. Zeng et al. [9] found that the maximum retrogradation and toughness of starch noodles occurred at −2 °C. In summary, the freezing temperature range is excessively wide and settings are inconsistent. This indicates that the scientific principle by which freezing transforms starch noodle texture from sticky paste to non-sticky, tough retrograded gel has not been fully clarified. Insufficient knowledge of the mechanism for optimizing freezing temperature to improve texture has seriously hindered understanding of temperature-mediated gel texture transformation in starch noodles. This knowledge gap leads to empirical process modifications characterized by blind adjustments and suboptimal results. The persistent issue of starch matrix adhesion continues to impair production efficiency and economic feasibility.

Because starch noodles are high-concentration starch gel (starch concentration > 35%) [3,5], controlling freezing temperature and duration is a crucial scientific issue both in practical application and for revealing the mechanistic relationship between freezing parameters and starch gel network evolution. The retrogradation of low-concentration starch gel (<20%) is dominated by amylose during short-term retrogradation; subsequently, interactions between starch chains strengthen during long-term retrogradation, while binding affinity between starch chains and water molecules weakens, leading to hardening [10], gel volume shrinkage, and water separation [11]. This property is utilized in sauces/dairy products to maintain liquidity [12]. However, high-concentration gels (>20%) contain more ordered molecular structures due to lower water content, higher density of starch molecules, and retained short-range ordered structure (e.g., single and double helices) [13], resulting in springiness loss and pulverization. This makes them suitable for texture setting in gel candies/plant-based foods [14]. To date, the extensive literature has focused on freezing temperature effects on low-concentration starch gels (<20% starch) [15,16], and retrogradation near T_g' promotes the formation of crystal nuclei [17,18]. The glass transition temperature (T_g') is a key kinetic parameter for multiscale food stability. When storage temperature is below T_g', molecular migration rates decrease by orders of magnitude, effectively inhibiting water migration, starch retrogradation, and oxidation reactions, thereby extending texture life of frozen/low-moisture foods. Modern food design recognizes T_g' regulation as a core strategy to extend shelf life and enhance functionality [19]. When freezing temperature is lower than T_g' [20–22], lower retrogradation results in a more viscous, softer, unformed, and fragile texture in starch noodles [1]. Conversely, soft bread and cake can delay staling and moisture loss by replacing 10–20% wheat flour with specific starches, maintaining lower retrogradation levels and extending shelf life by 3–5 days [23].

In summary, reasonable control of freezing temperature and time is the key to potato noodle processing, and the texture forming progress of high-concentrated starch gels frozen at below and above T_g' should be addressed. Therefore, this study aims to investigate the effects of different freezing temperatures (specifically T_5 (−18 °C) < T_4 (−12 °C) < T_3 (−6 °C) < T_g' < T_1 (freezing point)) on texture formation and surface compactness of

PSN, clarifying the key role of T_g' in this process. These findings will provide theoretical evidence and practical insights for optimizing freezing processes of starch gel-based foods through T_g' temperature control.

2. Results and Discussion

2.1. Determination of Freezing Curve and Glass Transition Temperature

The freezing curve reflects the variation in material temperature over time during the freezing process [24]. As shown in the freezing curve (Figure S1), the core temperature of the PSN reached freezing point at −1 °C. The supercooling point dropped to −4.5 °C, requiring 4 min to reach it. At this point, the liquid water begins to freeze, forming a crystal nucleus and releasing heat, causing a slight temperature increase. Subsequently, as freezing progresses, the temperature of the gel center decreases slowly [25]. Figure 1 shows the heat flow curve (Figure 1A) and its first derivative curve (Figure 1B) for PSN during DSC scanning (−60 °C to 60 °C). The arrow position on the first derivative curve indicates the glass transition temperature (T_g') of −3.1 °C [26]. It is reported that when starch gel water content exceeds 22%, T_g' is lower than room temperature, typically ranging between −3 °C and −10 °C [18]. Jiang et al. [27] report that as potato starch gel concentration increases (4%, 8%, 12%, and 16% (*w/w*)), freezing point temperatures decrease from −0.7 °C to −1.4 °C, while T_g' increases from −3.69 °C to −3.23 °C. In this study, the starch concentration was 35%, significantly higher than in Jiang et al. [27]. Consequently, the detected T_g' was higher than their values. Li et al. [28] studied the relationship between glass transition properties of frozen wheat dough and shiitake mushroom polysaccharide molecular weight. Their results showed that adding 2% high-MW (20 kDa) polysaccharide increased the dough glass transition temperature from −29.93 °C to −20.59 °C. Baek et al. [29] compared the effects of sugars/sugar alcohols (ribose, xylose, glucose, fructose, mannose, sucrose, maltose, isomaltose, trehalose, xylitol, mannitol, and sorbitol) on the thermal behavior and storage stability of corn starch gel (40% starch; starch/sugar = 10:1 or 10:3 *w/w*). They found that melting temperature and T_g' decrease with increasing sugar molar concentration.

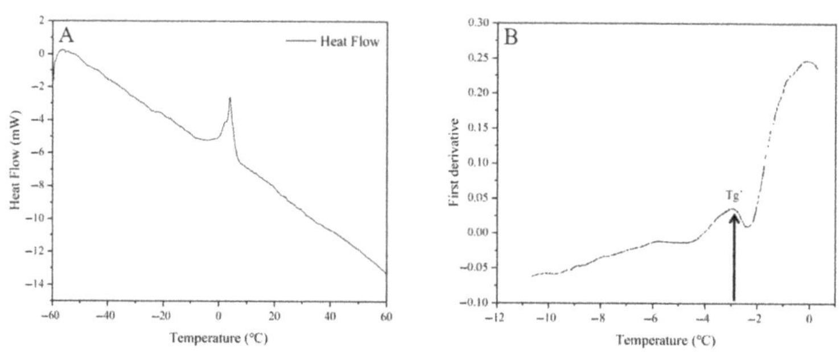

Figure 1. Heat flow curve (**A**) and first derivative of the heat flow curve from −11 °C to 0 °C (**B**) of potato starch gels.

2.2. Texture Characteristics of PSN

Hardness is defined as the force required to achieve a specific deformation [30]. As can be seen from Figure 2A, the hardness first reaches a steady state at −1 °C after 4 h, and then the upward trend gradually slows down. This is due to the appearance of a solid, delicate, and uniform gel network structure in the microstructure of the gel at this temperature [25,31,32]. The hardness at 3 °C and −6 °C increased from 0 h to 12 h, with no ice crystal formation during this process. Starch retrogradation formed a uniform gel structure. During the first rising stage, the hardness of PSN also reached its maximum value (13,734.096 g) at −3 °C for 12 h. Hardness decreased from 12 h to 24 h, likely due to ice crystal formation during water freezing, which disrupted the three-dimensional gel network [33]. However, between 24 h and 48 h, starch transitioned into a glassy state, increasing hardness [34]. Notably, long-term storage (>48 h) disrupts texture due to ice crystal growth [35], causing surface dehydration, void formation, brittle texture, and fracture [16]. Compared with −1, −3, and −6 °C, the hardness of −12 °C and −18 °C shows an overall upward trend, but the growth rate is significantly slower than −1, −3, and −6 °C (Figure 2A). This may occur when temperatures slightly higher than the glass transition temperature accelerate the retrogradation of starch gel and starch foods, and increase hardness and chewiness [36,37]. Simultaneously, the gel may enter a glassy state at extremely low temperatures, inhibiting retrogradation [27].

Adhesiveness is used to characterize the degree of cohesion in the internal structure of a sample [38]. The initial adhesiveness value of PSN was 80.373 $g{\cdot}s^{-1}$. After freezing at different temperatures, the adhesiveness significantly decreased with time (from 80.373 $g{\cdot}s^{-1}$ to 0 $g{\cdot}s^{-1}$). This indicates that freezing treatment can significantly reduce the adhesiveness of the PSN's surface. The adhesiveness of PSN after 4 h of freezing treatment at −12 °C and −18 °C (0 $g{\cdot}s^{-1}$ and 0.097 $g{\cdot}s^{-1}$, respectively) was close to 0 $g{\cdot}s^{-1}$. As the freezing time increased, the adhesiveness of PSN also tended to zero at −1, −3, and −6 °C. However, the trend of adhesiveness toward 0 at −12 °C and −18 °C was greater than that at −1, −3, and −6 °C. This may result from lower temperatures increasing water molecule mobility post thaw, reducing gel elasticity and smoothing the surface [21].

Springiness refers to the height ratio of a compressed deformed sample that returns to its pre-deformation condition after removing the deformation force [39]. Figure 2C shows springiness at −1, −3, and −6 °C increasing continuously from 0 to 12 h, peaking at 0.894 (−3 °C, 8 h), followed by a gradual decline. However, the temperatures of −12 °C and −18 °C increased from 0 h to 12 h. The maximum value is 0.915 (−18 °C, 8 h). Then, there was a continuous and slow downward trend. There was a small increase after 24 h. This indicated that within a certain range of freezing treatment, the springiness of PSN could be significantly improved. Previous studies suggest an optimal freezing range of −10 °C to −3 °C, where freezing is fastest. Temperatures outside this range reduce springiness and hardness [40].

Cohesiveness reflects intermolecular attraction strength and structural recovery after deformation [41]. As shown in Figure 2D, the cohesiveness fluctuates frequently, with the maximum value at −3 °C for 48 h (0.933) and the minimum value at −3 °C for 4 h (0.753). This may be because the potato starch gel has a faster recovery rate at −3 °C [25], intensifying competition between retrogradation and ice crystal formation. However, at −12 °C and −18 °C, cohesiveness stabilized and increased. Seetapan et al. [16] found that compared with a slow freezing rate, a higher freezing rate can better maintain the texture characteristics of starch gel samples with a total starch concentration of about 40%, because the water may undergo vitrification (or glassy ice formation) without any ice crystal formation.

Chewiness refers to the energy required to chew a solid sample into a stable state when swallowed [42], and is calculated from hardness, cohesiveness, and springiness. Chewiness exhibited similar trends with hardness at each freezing temperature, i.e., increasing initially, then decreasing within 24 h, with the maximum value occurring at −3 °C after 48 h (11,971.059 g). Values remained higher after 48 h, potentially due to glassy gel formation.

Resilience refers to the degree to which a sample can be restored to its original state after deformation under the same speed and pressure conditions [43]. Figure 2F shows resilience generally rising, then declining. Samples at −3 °C exhibited the greatest recovery, peaking at 0.835 units (−3 °C, 48 h) with a minimum of 0.622 units (−3 °C, 4 h). With the decline in freezing temperature (−6, −12, and −18 °C), the trend of first rising and then falling becomes stable.

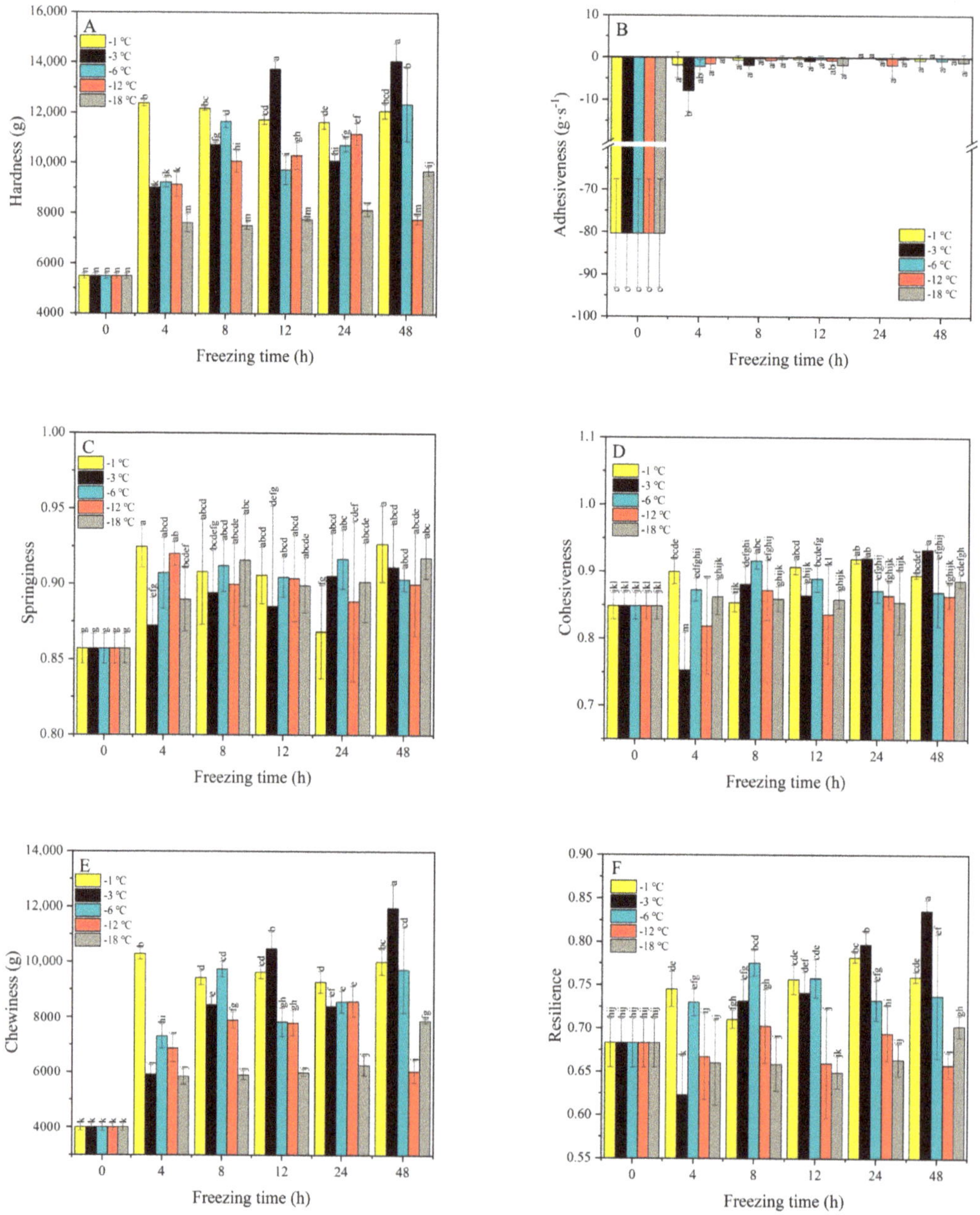

Figure 2. Effect of different freezing temperatures and times on textural properties of PSN (Textural properties: (**A**) Hardness, (**B**) Adhesiveness, (**C**) Springiness, (**D**) Cohesiveness, (**E**) Chewiness, (**F**) Resilience). The letters in the figure represent the differences between different groups.

2.3. Principal Coordinate Analysis (PCoA) and Cluster Analysis

To determine the decisive factors affecting texture dissimilarity among PSN samples, we evaluated the relative contributions of freezing time and temperature using six texture parameters (hardness, adhesiveness, springiness, cohesiveness, chewiness, resilience) via principal coordinate analysis (PCoA) and permutational multivariate analysis of variance (PERMANOVA) (Figure 3A,B). The first principal component primarily represents adhesiveness and springiness, whereas the second represents cohesiveness and resilience. As shown in Figure 3A, when freezing time was used as a variable to evaluate the similarity between PSN, samples were divided into four clusters: Cluster 1 (0 h), Cluster 2 (4 h), Cluster 3 (8 h and 24 h), and Cluster 4 (12 h and 48 h). This highlights significant intergroup differences in freezing time. Among them, PSN samples frozen for 8 h and 24 h, and 12 h and 48 h were clustered together, indicating that the texture of these treatments was similar. The texture of PSN samples frozen at 0 h and 4 h showed significant differences compared to other samples (p = 0.001). When freezing temperature was used as the variable, samples were separated into four groups, namely, Group 1 (−1 °C), Group 2 (−3 °C), Group 3 (−6 °C), and Group 4 (−12 °C and −18 °C), with significant differences (p = 0.001), confirming distinct textures across temperature groups. Notably, the R^2 value for time (R^2 = 0.963) exceeded that for temperature (R^2 = 0.947), indicating freezing time contributes more to PSN texture differences.

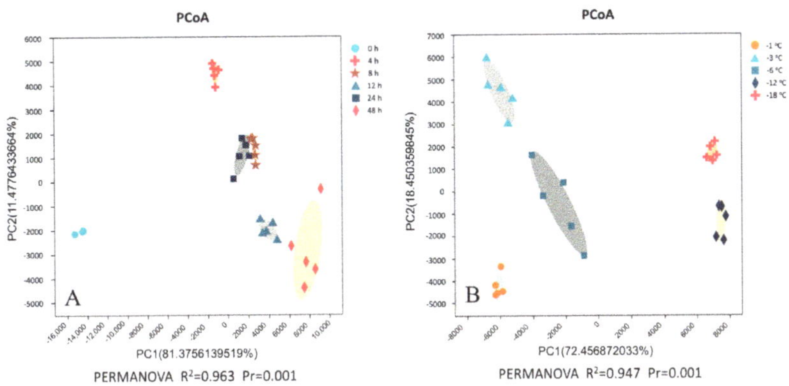

Figure 3. Principal coordinate analysis and PERMANOVA analysis based on textural data of PSN for different groups: freezing times (**A**) and freezing temperatures (**B**).

The hierarchical clustering analysis divides different clusters based on hardness, adhesiveness, springiness, cohesiveness, chewiness, and resilience. From the HCA plot in Figure 4A, Cluster 1 (blue cluster) represents the 0 h group; Cluster 2 (green cluster) includes the 4 h, 8 h, and 24 h groups; and Cluster 3 (red cluster) includes the 12 h and 48 h groups. These results are consistent with PCoA, emphasizing the quality changes of samples at different time points affected by starch recovery and ice crystal growth. From the HCA plot in Figure 4B, Cluster 1 (blue cluster) represents the −12 °C and −18 °C groups, while Cluster 2 (green cluster) includes the −3 °C groups, and Cluster 3 (red cluster) includes the −1 and −6 °C groups. This consistency with PCoA emphasizes texture differences between the −3 °C and adjacent groups. The results indicate that the optimal solidification temperature range for starch is between 10 °C and −3 °C. When the setting temperature of the same setting time is higher or lower than the temperature range, the physical properties of starch gel will change [40]. This HCA grouping confirms the texture features of PSN.

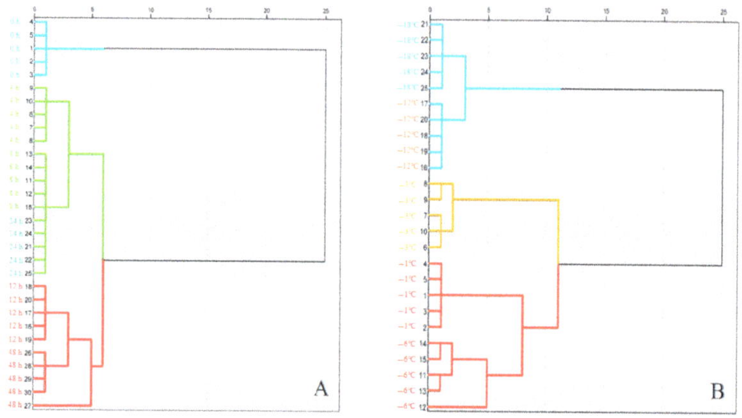

Figure 4. Dendrogram generated with freezing times (**A**) and freezing temperatures (**B**) based on hierarchical cluster analysis. (Different colors represent different cluster: blue represents cluster 1, green represents cluster 2, and red represents cluster 3).

2.4. Enzymatic Hydrolysis and SEM

According to the results of PCoA and cluster analysis, samples at −6 °C and below had a similar texture. Therefore, PSN samples frozen at −1, −3, and −12 °C were selected as representatives to observe the cross-section morphological characteristics. The starch gel before freezing presented (Figure 5A) a typical honeycomb-like gel structure after gelatinization. These loose pore structures provide the springiness and soft texture for starch noodles [44]. However, it was accompanied by high adhesiveness. After freezing for a certain time, a fissured texture appeared on the cross-section of the PSN. The results showed that starch retrogradation made the PSN gel structure compact and able to resist boiling water pressure. The typical structure of PSN is formed. This explained the rapid decline in adhesiveness after freezing in terms of microstructure.

At −1 °C (Figure 5B–D), due to being at the freezing point temperature, most of the water molecules close to the freezing point began to form some ice crystals, which put pressure on the structure of starch molecules and may have caused small cracks in some cases. At this temperature, the molecular chains can be partially crosslinked to form a network. However, because the retrogradation process of starch is inhibited due to the temperature approaching the freezing point, there is less opportunity for the rearrangement of molecules and the reformation of hydrogen bonds, which makes the physical properties of the gel remain stable [45], but it did not achieve the best retrogradation effect. It can be seen from the figure that the structure formed by 24 h (Figure 5C) of freezing was stronger than that formed by 4 h (Figure 5B) and 48 h (Figure 5D) of freezing, probably because the 24 h starch gel had the largest retrogradation degree. This also explained why the cohesiveness after 24 h of freezing was 0.92 g, which was significantly higher than that after 4 h of freezing (0.9 g) and 48 h of freezing (0.89). In addition, it can be clearly seen from Figure 5C-1 and D-1 that a large hole appeared at the center of the PSN section and the honeycomb structure became more blurred as it approached the edge of the section. This was because when the PSN were stored at a humidity lower than the refrigerator (RH 65.59% ± 0.62%), there was a moisture gradient between the PSN's surface and the freezing environment. In this case, the moisture in the PSN migrated from the interior to the surface, resulting in a decrease in moisture in the PSN during the initial freezing stage. On the other hand, the crosslinking degree of the macromolecular network structure in the PSN increases with the prolongation of freezing time. As a result, the transfer of water from the inside to the surface is inhibited. The transfer speed is lower than the diffusion speed of water on the PSN's surface [46,47].

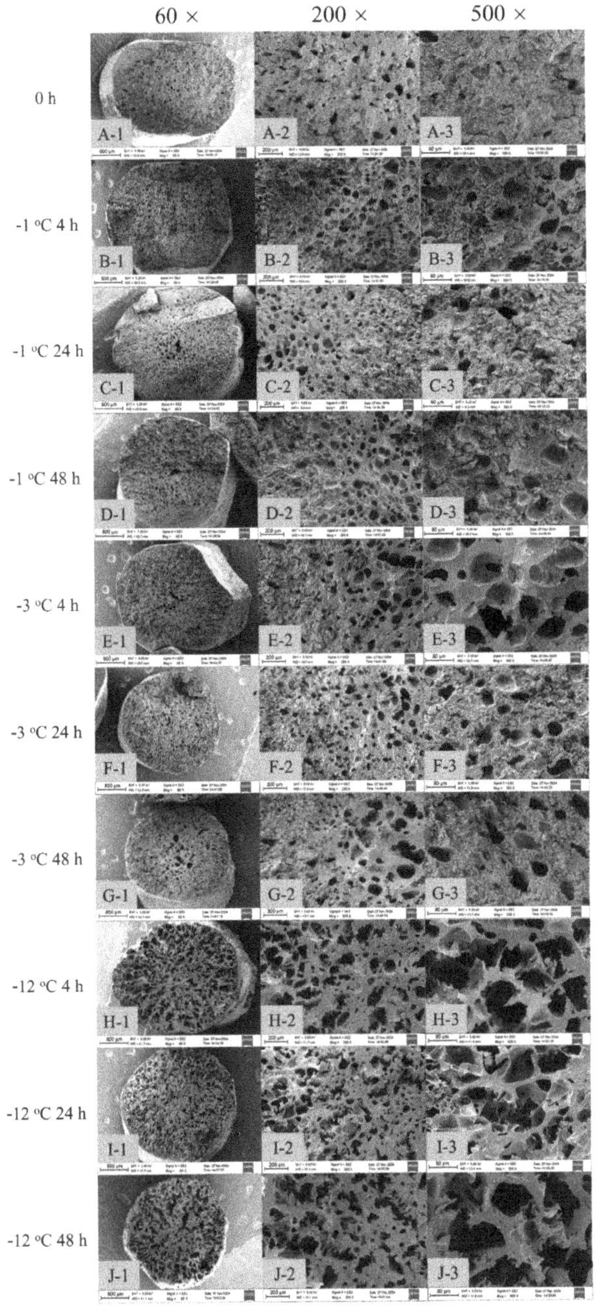

Figure 5. SEM images of unfrozen PSN (0 h) and frozen PSN sections at different freezing temperatures and times (Three groups of magnification (60×, 200×, 500×) freezing temperature time grouping: (**A**) 0 h, (**B**) −1 °C 4 h, (**C**) −1 °C 24 h, (**D**) −1 °C 48 h, (**E**) −3 °C 4 h, (**F**) −3 °C 24 h, (**G**) −3 °C 48 h, (**H**) −12 °C 4 h, (**I**) −12 °C 24 h, (**J**) −12 °C 48 h).

At −3 °C (Figure 5E–G), almost reaching the glass transition temperature, the water molecules in the gel were frozen. The thermal motion of starch molecules is relatively low, but it does not necessarily mean complete stillness [48]. The flexibility and fluidity of molecular chains have decreased, mainly manifested as vibration and small displacement [49]. At this time, hydrogen bond and other non-covalent interactions become more important, and the intermolecular interaction force increases, creating a more ordered structure (e.g., double helix, short-range molecular order), which helps to maintain the structural stability of gel [50]. This freezing phenomenon also led to changes in the mechanical properties of water molecules, especially affecting the hardness and resilience of the gel. It can be seen from the figure that the pore distribution formed by 24 h (Figure 5F) of freezing was more regular than that of 4 h (Figure 5E) and 48 h (Figure 5G) of freezing, which may have been because the 24 h starch gel had the largest retrogradation degree. This also explained why the cohesiveness value after 24 h of freezing was the highest, at 0.92 g, significantly higher than that after 4 h of freezing (0.75 g). At −12 °C (Figure 5H–J), due to the extremely low freezing temperature, the water contained in the gel formed ice nuclei at multiple points inside, which grew with the freezing time, destroying the structure of the gel that should have been continuously regenerated. As can be seen from the figure, the 24 h (Figure 5I) freezing had more fissured continuous structures than the 4 h (Figure 5H) and 48 h (Figure 5J) freezing, probably because the 24 h starch gel had the greatest degree of retrogradation. This also explained why freezing for 24 h had the highest cohesiveness value of 0.86, significantly higher than freezing for 4 h (0.82).

Figure 6 shows SEM images of PSN after 120 min enzymatic hydrolysis assessing surface firmness. In the 0 h group (Figure 6A) without freezing time and temperature treatment and only treated with α-amylase hydrolysis, the surface of the PSN was relatively uniformly hydrolyzed, showing a number of deeper and larger pores. This could be attributed to the gel-like texture of unfrozen noodles, giving more contact sites for amylase to degrade the gel to smaller fragments.

When frozen for 4 h, PSN samples stored at −1 °C (Figure 6B) and −3 °C (Figure 6E) had similar honeycomb surfaces, while samples frozen at −12 °C (Figure 6H) had cracked structures. The reason might be that the lower freezing temperature promoted the water inside the PSN to form ice crystals faster. The growth of ice crystals had a certain extrusion and destruction effect on the starch network structure of PSN, which led to the structural changes of the surface occurring more easily in the subsequent enzymatic hydrolysis process. When the freezing time reached 24 h, irregular pore structures appeared on the surface of PSN at −1 °C (Figure 6C) and −3 °C (Figure 6F), and the surface structure was more disordered at −1 °C compared to −3 °C. This might be because when the temperature dropped to −3 °C, the starch molecular chains underwent cooling, and some molecules would cross the energy barrier, forming temporary hydrogen bonds. As the temperature decreased, the phenomenon of retrogradation might be suppressed, but small-scale hydrogen bonding could lead to local structural stability. At −12 °C for 24 h (Figure 6I), the surface of the PSN almost completely lost its original smooth structure, showing a serious state of fragmentation and spalling. This showed that the structural damage of frozen PSN for 24 h became more and more serious with the decrease in temperature, which was because the ice crystals had more time to grow and gather, causing more serious damage to the starch network structure of PSN, resulting in serious damage to the surface structure in the process of enzymatic hydrolysis. Xu et al. [21] find that during refrigeration, the onset of ice crystal expansion destroys the short-range ordered structure of starch, resulting in an increase in pore size and structural relaxation of the submicroscopic structure, making the entire surface of PSN appear as irregular macropores and ruptures.

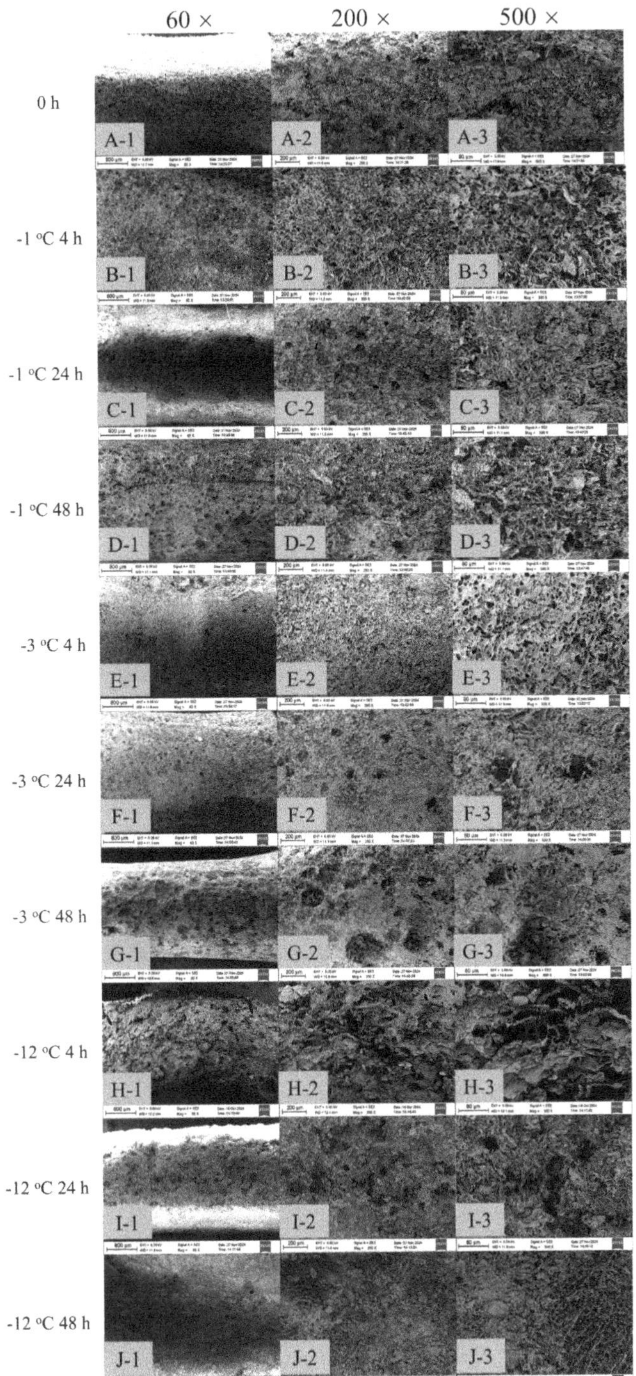

Figure 6. The surface SEM images of unfrozen (0 h) and frozen PSN after hydrolysis with 1% amylase for 120 min (Three groups of magnification (60×, 200×, 500×) freezing temperature time grouping: (**A**) 0 h, (**B**) −1 °C 4 h, (**C**) −1 °C 24 h, (**D**) −1 °C 48 h, (**E**) −3 °C 4 h, (**F**) −3 °C 24 h, (**G**) −3 °C 48 h, (**H**) −12 °C 4 h, (**I**) −12 °C 24 h, (**J**) −12 °C 48 h).

When the freezing time reached 48 h, the degree of damage on the surface of the PSN was further aggravated. At −1 °C for 48 h (Figure 6D), most areas were broken and peeled. At −3 °C for 48 h (Figure 6G), the PSN had basically lost their original shape, and the surface showed collapse and fragmentation, and the original structure of the PSN could hardly be recognized. At −12 °C for 48 h (Figure 6J), the PSN showed a uniform but slightly hydrolyzed surface morphology, indicating a tighter and firmer surface possibly squeezed by internal ice.

3. Conclusions

This study provides a detailed analysis of the impact of freezing temperature on the texture formation of PSN. The results demonstrate that the freezing conditions significantly influence the mechanical and microstructural properties of PSN, with the most desirable texture being achieved when PSN are frozen at −3 °C for 24 h. Freezing below T_g' (at −6, −12, and −18 °C) resulted in a porous structure disrupted by the growth and extrusion of ice crystals and lower springiness, while temperatures around the freezing point (−1 °C) and T_g' (−3 °C) produced noodles with better texture, tighter surface, and continuous solid cross-sections. These improvements are attributed to the starch retrogradation, reduced ice crystal formation, and higher mobility of starch chains at these temperatures. This study highlights the importance of T_g' in the texture formation of starch gel during freezing and provides valuable insights for optimizing freezing processes in starch-based food products. Further research should focus on the mechanisms behind the texture changes of starch gels at temperatures below and above T_g' from the molecular view.

4. Materials and Methods

4.1. *Materials*

Potato starch: S11004, Shanghai Yuanye Bio-Technology Co., Ltd. (Shanghai, China), containing 17.14% moisture, 0.21% ash, 0.15% protein, 0.36% lipid, and 34.99% amylose. α-Amylase (*Bacillus subtilis*): S10002, Shanghai Yuanye Technology Co., Ltd. (Shanghai, China), enzyme activity 4000 u/g.

4.2. *Methodology*

4.2.1. Preparation of PSN

Binder potato starch (20 g) was dispersed in 120 mL of water and heated in a boiling water bath for 2 min with continuous stirring to yield a totally gelatinized paste. The resultant paste was mixed with 200 g of potato starch and kneaded into a uniform and smooth starch dough. The kneaded smooth dough was placed in the artificial extrusion equipment as to apply uniform pressure on the starch powder and squeeze it into the boiling water pot (the distance between the artificial extrusion equipment and the pot should be kept at 30 cm). After being cooked for 1 min, noodles were transferred into cold water for cooling for 5 min and dried at room temperature for 2 h.

4.2.2. Measurement of Freezing Point Temperature of PSN

The sensor of the contact thermometer (DT1310, Shenzhen Lihua Instrument & Tools Co., Ltd., Shenzhen, China) was inserted into the starch gel center, and the gel was placed in a cryogenic refrigerator (BCD-561WLHSS14W9U1, Haier Group, Shandong, China) at −24 °C for quick freezing. The core temperature of the gel was recorded every 2 min. The freezing curve was drawn with time as the abscissa and the gel core temperature as the ordinate [25].

4.2.3. Measurement of Glass Transition Temperature of PSN

The glass transition temperature of the potato starch gel sample was measured using a differential scanning calorimeter (DSC, DSC8000, PerkinElmer, Shelton, CT, USA). The sample powder (approximately 3 mg) was accurately weighed into a 40 μL DSC aluminum crucible, and distilled water was added to achieve a starch-to-water ratio of 1:3 (*w*/*v*). Prior to measurement, the aluminum crucible was sealed and equilibrated overnight at room temperature. An empty aluminum crucible was used as a reference. The sample was scanned across the temperature range of −60 °C to 60 °C with a heating rate of 5 °C/min. The thermograms obtained from the heating cycle were used to determine the glass transition temperature ($T_g{}'$) using Universal Analysis software (Universal Analysis 2000) of TA Instruments [51].

4.3. Freezing of PSN

PSN were frozen in six freezing temperature gradients, T_5 (−18 °C) < T_4 (−12 °C), T_3 (−6 °C) < $T_g{}'$ (glass transition temperature) < T_1 (freezing point temperature) for 48 h, and samples were taken at 0, 4, 8, 12, 24, and 48 h for subsequent experiments. Blank samples were not subjected to freezing treatment and were recorded as 0 h.

4.3.1. Textural Properties

A texture analyzer (TA-XT2I texture analyzer, Stable Micro Systems, Godalming, UK) was used to determine the parameters. The HDP/PFS probe was selected, and five PSN with no cracking, no bending, uniform thickness, and the same length were selected. Samples were thawed for 10 min. Subsequently, the HDP/PFS probe was used for TPA testing. During testing, surface moisture was removed by blotting with absorbent paper, and five PSN strips were placed on the test bench in parallel. This was performed at least 20 times in parallel experiments on each sample. The result was processed by removing the maximum and minimum values and averaging them. The parameters were set as follows: measurement mode: compression mode; speed before test: 1.00 mm/s; operating speed: 0.5 mm/s; speed after test: 10.00 mm/s; target mode: flexibility; flexibility: 40%; trigger type: automatic; trigger force: 3.0 g [52–54].

4.3.2. Enzymatic Hydrolysis and Scanning Electron Microscope (SEM)

The surface microstructure of PSN before and after enzymatic hydrolysis for 120 min was observed using an SEM (SUPRATM55, Carl Zeiss AG, Oberkochen, Germany). An enzymatic solution with a pH value of 6.0 and a dry matter mass of 1% was prepared. The enzymatic hydrolysis temperature was maintained at 60 °C, and samples were cut to 2 cm, placed in the enzymatic hydrolysis solution, and subjected to continuous stirring at 150 r/min. Enzymatic hydrolysis was carried out for 120 min [55]. Subsequently, samples were removed for adhesive fixation, and the surface and cross-section of the samples were gold-coated using the sputtering coating method. Scanning electron microscopy was performed at an acceleration voltage of 15 kV, and photographs were taken; the PSN surface images were scanned at a magnification of ×50 [56].

4.4. Statistical Analysis

Data were processed in WPS office (Kingsoft, WPS Office 12.1.0.21915, Beijing, China); statistical analysis and cluster analysis (intergroup connection method, squared Euclidean distance) in this study were both based on IMB SPSS Statistics 27 Free Edition (SPSS, Chicago, IL, USA). Principal component analysis (Euclidean algorithm) was conducted on the Majorbio platform (https://www.majorbio.com/, URL (accessed on 6 June 2025)), and plotted data were processed by Origin 2021 (Origin-Lab, Inc., Northampton, MA, USA).

The significant differences were checked by one-way ANOVA (Duncan's test), and the value of $p < 0.05$ was deemed statistically significant. All experiments were replicated five times and data were presented as mean ± standard deviation.

Supplementary Materials: The following supporting information can be downloaded at https://www.mdpi.com/article/10.3390/gels11080639/s1: Figure S1: Temperature change in the center of PSN.

Author Contributions: Conceptualization, H.L. and S.Y.; methodology, H.L. and S.Y.; software, H.L.; validation, H.L.; formal analysis, H.L.; investigation, H.L. and Q.H.; resources, S.Y., L.L. and X.D.; data curation, H.L.; writing—original draft preparation, H.L.; writing—review and editing, S.Y. and L.L.; visualization, H.L.; supervision, S.Y., L.L. and X.D.; project administration, S.Y. and L.L. All authors have read and agreed to the published version of the manuscript.

Funding: This research was supported by the National Natural Science Foundation of China (No. 32402159); the Key R&D Program of Shandong Province, China (2024CXGC010917); and the Agricultural Science and Technology Innovation Project of Shandong Academy of Agricultural Sciences (CXGC2025H18).

Institutional Review Board Statement: Not applicable.

Informed Consent Statement: Not applicable.

Data Availability Statement: Data are contained within the article.

Conflicts of Interest: The authors declare no conflicts of interest.

Abbreviations

The following abbreviations are used in this manuscript:

PSN	Potato starch noodles
PCoA	Principal coordinate analysis
SEM	Scanning electron microscope

References

1. Tan, H.; Li, Z.G.; Tan, B. Starch noodles: History, classification, materials, processing, structure, nutrition, quality evaluating and improving. *Food Res. Int.* **2009**, *42*, 551–576. [CrossRef]
2. Wang, S.; Chen, Z.; Zheng, G. Effects of the addition of starches in wheat flour on fried instant noodle quality. *Food Sci.* **2005**, *11*, 109–111.
3. Xiang, Z.; Ye, F.; Zhou, Y.; Wang, L.; Zhao, G. Performance and mechanism of an innovative humidity-controlled hot-air drying method for concentrated starch gels: A case of sweet potato starch noodles. *Food Chem.* **2018**, *269*, 193–201. [CrossRef]
4. Sun, L.; Zhu, Z.; Sun, D.-W. Regulating ice formation for enhancing frozen food quality: Materials, mechanisms and challenges. *Trends Food Sci. Technol.* **2023**, *139*, 104116. [CrossRef]
5. Yang, S.; Zhang, M.; Shan, C.; Chen, Z. Evaluation of cooking performance, structural properties, storage stability and shelf life prediction of high-moisture wet starch noodles. *Food Chem.* **2021**, *357*, 129744. [CrossRef] [PubMed]
6. Fan, H.; Li, R.; Wang, N.; Yang, H.; Zhao, T.; Li, N.; Ai, Z. Effect of freezing process on the quality of sweet potato vermicelli without aluminum. *Food Ferment. Ind.* **2016**, *42*, 142–146. [CrossRef]
7. Sun, Z.; Mu, T.; Ma, M.; Chen, J. Processing technology optimization and physicochemical properties of alum-free fresh potato/cassava starch noodles. *Mod. Food Sci. Technol.* **2020**, *36*, 153–160. [CrossRef]
8. Nguyen, T.P.; Rumpagaporn, P.; Songsermpong, S. Investigation of freezing temperature and time to improve resistant starch content and quality of pure mung bean starch vermicelli. *Aust. J. Crop Sci.* **2023**, *17*, 693–698. [CrossRef]
9. Zeng, J.; Jiang, J.; Gao, H.; Meng, K.; Song, M.; Li, G. Optimization of the Aging Process for Potato Vermicelli without Additives and Quality Changes during Storage. *Food Sci.* **2019**, *40*, 283–289. [CrossRef]
10. Miles, M.J.; Morris, V.J.; Orford, P.D.; Ring, S.G. The roles of amylose and amylopectin in the gelation and retrogradation of starch. *Carbohydr. Res.* **1985**, *135*, 271–281. [CrossRef]
11. Keetels, C.J.A.M.; Vliet, T.v.; Walstra, P. Gelation and retrogradation of concentrated starch systems: 2. Retrogradation. *Food Hydrocoll.* **1996**, *10*, 355–362. [CrossRef]

12. Sikora, M.; Kowalski, S.; Tomasik, P.; Sady, M. Rheological and sensory properties of dessert sauces thickened by starch–xanthan gum combinations. *J. Food Eng.* **2007**, *79*, 1144–1151. [CrossRef]
13. Ring, S.G.; Colonna, P.; I'Anson, K.J.; Kalichevsky, M.T.; Miles, M.J.; Morris, V.J.; Orford, P.D. The gelation and crystallisation of amylopectin. *Carbohydr. Res.* **1987**, *162*, 277–293. [CrossRef]
14. Burey, P.; Bhandari, B.R.; Howes, T.; Gidley, M.J. Hydrocolloid gel particles: Formation, characterization, and application. *Food Sci. Nutr.* **2008**, *48*, 361–377. [CrossRef]
15. Hu, Y.; Li, C.; Tan, Y.; McClements, D.J.; Wang, L. Insight of rheology, water distribution and in vitro digestive behavior of starch based-emulsion gel: Impact of potato starch concentration. *Food Hydrocoll.* **2022**, *132*, 107859. [CrossRef]
16. Seetapan, N.; Limparyoon, N.; Gamonpilas, C.; Methacanon, P.; Fuongfuchat, A. Effect of cryogenic freezing on textural properties and microstructure of rice flour/tapioca starch blend gel. *J. Food Eng.* **2015**, *151*, 51–59. [CrossRef]
17. Moo, Y.B.; Kwang, J.K.; Ki, C.C.; Yeon, C.H.; Wang, S.K. Recrystallization kinetics and glass transition of rice starch gel system. *Food Chem.* **1997**, *45*, 4242–4248. [CrossRef]
18. Tananuwong, K.; Reid, D.S. Differential scanning calorimetry study of glass transition in frozen starch gels. *J. Agric. Food Chem.* **2004**, *52*, 4308–4317. [CrossRef] [PubMed]
19. Dominique, C.; Martine, L.M.; Denise, S. Towards an improved understanding of glass transition and relaxations in foods: Molecular mobility in the glass transition range. *Trends Food Sci. Technol.* **2000**, *11*, 41–55. [CrossRef]
20. Charoenrein, S.; Preechathammawong, N. Undercooling associated with slow freezing and its influence on the microstructure and properties of rice starch gels. *J. Food Eng.* **2010**, *100*, 310–314. [CrossRef]
21. Xu, M.; Ji, S.; Ren, Y.; Xu, Y.; Zhao, X.; Zhou, Z.; Wang, Y.; Li, Y.; Shen, J.; Li, K.; et al. Enhancing 3D printing accuracy of cassava starch gel through temperature-controlled freezing. *Food Biosci.* **2024**, *62*, 105526. [CrossRef]
22. Zhu, B.; Zhan, J.; Chen, L.; Tian, Y. Amylose crystal seeds: Preparation and their effect on starch retrogradation. *Food Hydrocoll.* **2020**, *105*, 105805. [CrossRef]
23. Hayes, A.M.R.; Okoniewska, M.; Martinez, M.M.; Zhao, B.; Hamaker, B.R. Investigating the potential of slow-retrograding starches to reduce staling in soft savory bread and sweet cake model systems. *Food Res. Int.* **2020**, *138*, 109745. [CrossRef]
24. Sun, X.; Liu, W.; Xiao, H.; Liu, C.; Yang, F.; Hu, J.; Wang, Y.; Zhao, R.; Wang, H. Effects of different freezing methods on the damage in potato tissues impregnated with trehalose. *LWT* **2024**, *213*, 117096. [CrossRef]
25. Jiang, J.; Zeng, J.; Gao, H.; Zhang, L.; Wang, F.; Su, T.; Xiang, F.; Li, G. Effect of low temperature on the aging characteristics of a potato starch gel. *Int. J. Biol. Macromol.* **2020**, *150*, 519–527. [CrossRef]
26. Jeong, D.; Lee, J.H.; Chung, H.J. Effect of molecular structure on phase transition behavior of rice starch with different amylose contents. *Carbohydr. Polym.* **2021**, *259*, 117712. [CrossRef] [PubMed]
27. Jiang, J.; Gao, H.; Zeng, J.; Zhang, L.; Wang, F.; Su, T.; Li, G. Determination of subfreezing temperature and gel retrogradation characteristics of potato starch gel. *LWT* **2021**, *149*, 112037. [CrossRef]
28. Li, S.; Sui, W.; Cui, J.; Wan, Z.; Zhang, J.; Wu, T.; Liu, D.; Zhang, M. Glass transition features of frozen dough regulated by lentinan molecular weight and their correlation with dough structural and rheological properties. *Food Hydrocoll.* **2025**, *164*, 111150. [CrossRef]
29. Baek, M.H.; Yoo, B.; Lim, S.T. Effects of sugars and sugar alcohols on thermal transition and cold stability of corn starch gel. *Food Hydrocoll.* **2004**, *18*, 133–142. [CrossRef]
30. Han, N.-R.; Bae, J.-E.; An, H.W.; Yun, H.-J.; Retnoaji, B.; Lee, S.; Lee, S.G. Influence of potato starch on the 3D printing of senior-friendly foods enriched with oyster powder. *LWT* **2025**, *224*, 117886. [CrossRef]
31. Jang, J.K.; Pyun, Y.R. Effect of moisture level on the crystallinity of wheat starch aged at different temperatures. *Starch-Stärke* **2006**, *49*, 272–277. [CrossRef]
32. Bárcenas, M.E.; Haros, M.; Benedito, C.; Rosell, C.M. Effect of freezing and frozen storage on the staling of part-baked bread. *Food Res. Int.* **2003**, *36*, 863–869. [CrossRef]
33. Seetapan, N.; Limparyoon, N.; Fuongfuchat, A.; Gamonpilas, C.; Methacanon, P. Effect of freezing rate and starch granular morphology on ice formation and non-freezable water content of flour and starch gels. *Int. J. Food Prop.* **2015**, *19*, 1616–1630. [CrossRef]
34. Goff, H.D. Low-temperature stability and the glassy state in frozen foods. *Food Res. Int.* **1992**, *25*, 317–325. [CrossRef]
35. Zou, Z.; Li, M.; Song, Q.; Guo, B.; Zhou, G.; Xie, F. Regulation of retrogradation and textural properties of mung bean starch gels during storage: Role of cooling rate. *Carbohydr. Polym.* **2025**, *366*, 123915. [CrossRef]
36. Czuchajowska, Z.; Otto, T.; Paszczynska, B.; Baik, B.K. Composition, thermal behavior, and gel texture of prime and tailings starches from garbanzo beans and peas. *Cereal Chem.* **1998**, *75*, 466–472. [CrossRef]
37. Qiao, K.; Peng, B. Freezing rate's impact on starch retrogradation, ice recrystallization, and quality of water-added and water-free quick-frozen rice noodles. *Int. J. Biol. Macromol.* **2024**, *276*, 134047. [CrossRef]
38. Lee, T.-H.; Park, J.-W. Assessing adhesiveness levels in a dysphagia diet for older adults. *Geriatrics* **2024**, *9*, 48. [CrossRef]

39. Faber, T.J.; Jaishankar, A.; McKinley, G.H. Describing the firmness, springiness and rubberiness of food gels using fractional calculus. Part I: Theoretical framework. *Food Hydrocoll.* **2017**, *62*, 311–324. [CrossRef]
40. Hager, A.-S.; Wolter, A.; Czerny, M.; Bez, J.; Zannini, E.; Arendt, E.K.; Czerny, M. Investigation of product quality, sensory profile and ultrastructure of breads made from a range of commercial gluten-free flours compared to their wheat counterparts. *Eur. Food Res. Technol.* **2012**, *235*, 333–344. [CrossRef]
41. Hadde, E.K.; Chen, J. Shear and extensional rheological characterization of thickened fluid for dysphagia management. *J. Food Eng.* **2019**, *245*, 18–23. [CrossRef]
42. Cheng, Z.; Zhang, B.; Qiao, D.; Yan, X.; Zhao, S.; Jia, C.; Niu, M.; Xu, Y. Addition of κ-carrageenan increases the strength and chewiness of gelatin-based composite gel. *Food Hydrocoll.* **2022**, *128*, 107565. [CrossRef]
43. Koekuyt, H.A.; Dobson, S.; Marangoni, A.G. Lipid complexation improves the mechanical properties and functionality of legume starch gels. *Food Hydrocoll.* **2025**, *167*, 111401. [CrossRef]
44. Liu, J.; Zhang, R.; Pi, X.; Zhang, B. Structural features of rice starch-protein system: Influence of retrogradation time and quick-freezing temperature. *Int. J. Biol. Macromol.* **2024**, *277*, 133981. [CrossRef] [PubMed]
45. Park, E.Y.; Baik, B.-K.; Lim, S.-T. Influences of temperature-cycled storage on retrogradation and in vitro digestibility of waxy maize starch gel. *J. Cereal Sci.* **2009**, *50*, 43–48. [CrossRef]
46. Zhu, T.; Yang, J.; Qin, W.; Tian, Y.; Wang, Y.; Guo, X. Drying kinetics of soy protein isolate-corn starch film during preparation and its moisture adsorption characteristics during storage. *Grain Oil Sci. Technol.* **2023**, *6*, 120–126. [CrossRef]
47. Ichikawa, K.; Mori, T.; Kitano, H.; Fukuda, M.; Mochizuki, A.; Tanaka, M. Fourier transform infrared study on the sorption of water to various kinds of polymer thin films. *J. Polym. Sci.* **2001**, *39*, 2175–2182. [CrossRef]
48. Hoover, R. Composition, molecular structure, and physicochemical properties of tuber and root starches: A review. *Carbohydr. Polym.* **2001**, *45*, 253–267. [CrossRef]
49. Wu, X.; Sun, H.; Qin, Z.; Che, P.; Yi, X.; Yu, Q.; Zhang, H.; Sun, X.; Yao, F.; Li, J. Fully physically crosslinked pectin-based hydrogel with high stretchability and toughness for biomedical application. *Int. J. Biol. Macromol.* **2020**, *149*, 707–716. [CrossRef]
50. McGrane, S.J.; Mainwaring, D.E.; Cornell, H.J.; Rix, C.J. The role of hydrogen bonding in amylose gelation. *Starch-Stärke* **2004**, *56*, 122–131. [CrossRef]
51. Chen, L.; Ren, F.; Zhang, Z.; Tong, Q.; Rashed, M.M.A. Effect of pullulan on the short-term and long-term retrogradation of rice starch. *Carbohydr. Polym.* **2015**, *115*, 415–421. [CrossRef] [PubMed]
52. Yi, C.; Zhu, H.; Bao, J.; Quan, K.; Yang, R. The texture of fresh rice noodles as affected by the physicochemical properties and starch fine structure of aged paddy. *LWT* **2020**, *130*, 109610. [CrossRef]
53. Yan, X.; Wu, J.; Zhao, C.; Luo, S.; Huang, L.; Guo, D.; Liu, C. Chinese rice noodles form the viscoelastic texture by dual high-temperature retrogradation: An insight into the mechanism. *LWT* **2023**, *189*, 115496. [CrossRef]
54. Liu, X.; Chao, C.; Yu, J.; Copeland, L.; Wang, S. Mechanistic studies of starch retrogradation and its effects on starch gel properties. *Food Hydrocoll.* **2021**, *120*, 106914. [CrossRef]
55. Cao, H.; Huang, M.; Liu, Z.; Huang, K.; Wang, M.; Song, H.; Li, S.; Zhang, Y.; Sun, Z.; Guan, X. Influencing the retrogradation properties of potato mash through multi-scale structural modifications mediated by enzymatic hydrolysis forms. *Food Chem.* **2025**, *485*, 144467. [CrossRef]
56. Ma, M.; Zhang, X.; Zhu, Y.; Li, Z.; Sui, Z.; Corke, H. Mechanistic insights into the enhanced texture of potato noodles by incorporation of small granule starches. *Int. J. Biol. Macromol.* **2024**, *257*, 128535. [CrossRef] [PubMed]

gels

MDPI

Article

Polydextrose Addition Improves the Chewiness and Extended Shelf-Life of Chinese Steamed Bread Through the Formation of a Sticky, Elastic Network Structure

Chang Liu [1,2], Bing Dai [1,3], Xiaohong Luo [1], Hongdong Song [2] and Xingjun Li [1,2,*]

1 Academy of National Food and Strategic Reserves Administration, National Engineering Research Center for Grain Storage and Transportation, Beijing 102209, China; shlglc@126.com (C.L.); bingdai3@126.com (B.D.); lxh@ags.ac.cn (X.L.)
2 College of Health Science and Engineering, University of Shanghai for Science and Technology, Shanghai 200093, China; cau4080@163.com
3 College of Grain and Strategic Reserves, Henan University of Technology, Zhengzhou 450001, China
* Correspondence: lxj@ags.ac.cn

Abstract: This study explored the effects of adding a newly developed type of polydextrose on the appearance, sensory score, and textural parameters of steamed bread and the microstructure of dough, as well as the pasting, thermal, and thermal mechanical properties of high-gluten wheat flours. The results revealed that, compared with a control sample, 3–10% of polydextrose addition significantly increased the hardness, adhesiveness, gumminess, and chewiness of steamed bread, but other textural parameters like springiness, cohesiveness, and resilience remained basically the same. Further, in contrast to the control sample, 3–10% polydextrose addition significantly reduced the specific volume and width/height ratio of steamed bread but increased the brightness index, yellowish color, and color difference; improved the internal structure; and maintained the other sensory parameters and total score. Polydextrose addition decreased the peak, trough, final, breakdown, and setback viscosity of the pasting of wheat flour suspension solutions but increased the pasting temperature. Polydextrose additions significantly reduced the enthalpy of gelatinization and the aging rate of flour paste but increased the peak temperature of gelatinization. A Mixolab revealed that, with increases in the amount of added polydextrose, the dough's development time and heating rate increased, but the proteins weakened, and the peak torque of gelatinization, starch breakdown, and starch setback torque all decreased. Polydextrose additions increased the crystalline regions of starch, the interaction between proteins and starch, and the β-sheet percentage of wheat dough without yeast and of steamed bread. The amorphous regions of starch were increased in dough through adding polydextrose, but they were decreased in steamed bread. Further, 3–10%of polydextrose addition decreased the random coils, α-helixes, and β-turns in dough, but the 3–7% polydextrose addition maintained or increased these conformations in steamed bread, while 10% polydextrose decreased them. In unfermented dough, as a hydrogel, the 5–7% polydextrose addition resulted in the formation of a continuous three-dimensional network structure with certain adhesiveness and elasticity, with increases in the porosity and gas-holding capacity of the product. Moreover, the 10% polydextrose addition further increased the viscosity, freshness, and looseness of the dough, with smaller and more numerous holes and indistinct boundaries between starch granules. These results indicate that the 3–10% polydextrose addition increases the chewiness and freshness of steamed bread by improving the gluten network structure. This study will promote the addition of polydextrose in steamed bread to improve shelf-life and dietary fiber contents.

 https://doi.org/10.3390/gels11070545

Keywords: polydextrose; Chinese steamed bread; continuous three-dimensional network structure; freshness and looseness; dough development time; water soluble dietary fiber

1. Introduction

A report on the nutrition and chronic disease status of Chinese residents indicated that in 2014, the number of overweight and obese people in China reached 225 million and 125 million, respectively, while there were 160 million hypertensive patients, 160 million people with lipid disorder, and 94 million individuals with abnormal blood sugar (diabetes) [1,2]. In 2019, the premature mortality rate among patients with major chronic diseases, including cardiovascular and cerebrovascular diseases, cancer, chronic respiratory diseases, and diabetes, in China was 16.5%, 2% lower than in 2015 [3]. The Report on the Monitoring of Chronic Diseases in the National Plan, published in March 2024, indicated that there were more than 480 million chronic patients in China in 2023, accounting for 30.5% of the total population [4]. The main cause is a lack of dietary fiber intake by residents [5]. Fiber is the indigestible component of plants. It does not provide any nutrients to the body, but it is essential for maintaining the healthy function of the human digestive system [6,7]. Cereal dietary fibers can prevent and alleviate diabetes, plasma cholesterol, cardiovascular diseases, rectal cancer, emphysema, and obesity [8,9]. Whole wheat steamed bread, with a dietary fiber content of 10%, has a very poor taste, but delicious white steamed bread only has about 1%dietary fiber [10,11]. White steamed bread is a staple traditional fermented food in China and Asia that is consumed by a large proportion of residents [12,13]. Thus, it is worth studying methods to increase the dietary fiber in steamed white bread through the addition of water-soluble dietary fibers.

The physical properties of fiber-containing foods allow them to regulate the colon, in part because of their ability to bind to water [14]. Polydextrose has a strong water-holding capacity [15]. Polydextrose is a highly branched polymer with randomly distributed α- and β-glycosidic bonds. It is synthesized with D-glucose in a reactor at a ratio of 89:10:1 (D-glucose–sorbitol–citric acid) [16]. Polydextrose has been used in the food industry in developed countries as a bulking agent and a low-energy ingredient, replacing sugar and fat since 1980 [17,18]. China introduced and synthesized this product in the year 2000. China produced 101,000 tons of polydextrose in 2020, with an annual output value of around USD 0.137 billion; production in 2026 is expected to reach 135,000 tons [19]. The use of polydextrose in health foods, beverages, fermented dairy products, and baked foods accounts for 39%, 21.6%, 16.6%, and 11.8%, respectively, of the total use of polydextrose in China, and the remaining 11% is used in chocolate, nutrition bars, and other foods [19]. Polydextrose is good for human health as a soluble dietary fiber [20]. Compared with insoluble dietary fiber, polydextrose has many more health and processing benefits. Polydextrose can be widely utilized in various foods, including high-fiber, low-energy functional foods, because it has fewer calories (1 cal/g), better stability, and high tolerance to sugars [21]. Polydextrose can also be used as a humectant in baked foods; it can delay moisture evaporation, prevent the product from spoiling, and extend shelf-life [22]. Polydextrose has excellent water-binding properties that help stabilize gluten [23] or bread dough [24] during freezing and thawing processes. Jing et al. [10] reported that a 1–9% addition of polydextrose as a soluble fiber had little effect on the sensory attributes of steamed bread. However, there is a lack of further research on the addition of polydextrose to steamed bread during dough preparation and the resulting effects on the freshness of the cooked steamed bread.

Steamed bread, made by steaming fermented dough, has a history of more than 1700 years. It is a popular food in China and many Asian countries [25]. However, there are differences in the dietary structure of steamed bread compared to whole wheat flour itself, and steamed bread contains low levels of lysine, vitamins, minerals, and dietary fiber; its nutritional components are not comprehensive nor balanced [25,26]. In this study, given the high moisture absorption rate of amorphous polydextrose powder, a new type of polydextrose (PD-ST) with high market cost performance was used, and 3–10% this polydextrose was added as the soluble fiber during the dough preparation of white steamed bread; the appearance, tasting score, and textural parameters of the steamed bread, as well as its gelatinization parameters and thermo-mechanical properties, were analyzed. The crystal conformation of starch and the secondary structure of protein in dough and steamed bread were comprehensively analyzed, and the microstructure of the unfermented dough was observed. The aim of this study was to improve the shelf-life, chewiness quality, and dietary fiber content of fermented steamed bread through the addition of polydextrose, thereby preventing and alleviating the incidence of chronic diseases in residents who mainly eat steamed bread.

2. Results and Discussion

2.1. Comparison of New Polydextrose and Conventional Polydextrose

New polydextrose (PD-ST), with 1.0% moisture content (MC), consists of irregular and inhomogeneous particles (Figure 1A), and it is different from conventional spherical polydextrose with protrusions (PD-XG, 4.7% MC) (Figure 1B). PD-ST has a higher moisture absorption rate than PD-XG over a 12 h measurement period at 26 °C and 84.1% relative humidity (Figure 2).

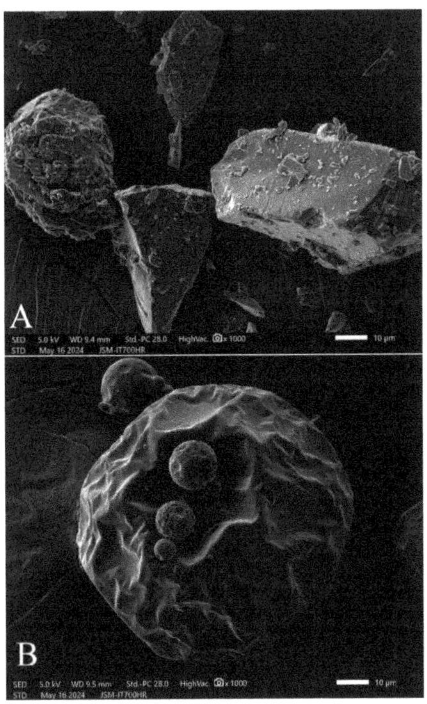

Figure 1. New polydextrose (PD-ST) and conventional polydextrose (PD-XG). Note: (**A**) PD-ST; (**B**) PD-XG. The photos are enlarged by1000×.

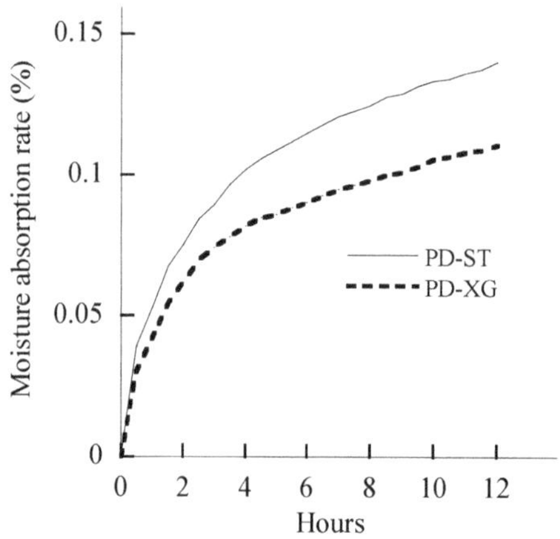

Figure 2. The moisture absorption rate of PD-ST and PD-XG polydextrose. Note: The data are expressed as mean; number of repetitions—n = 2.

Two polydextrose samples had at least 11 infrared spectroscopy absorption peaks (Figure 3A), and the peaks with an absorbance value of greater than 1 were 1027 and 3416 cm^{-1}, representing amorphous structure (primary alcohol C-OH) and hydroxyl functional groups, respectively (Figure 3B). This is a significant feature of amorphous-phase polydextrose powder. The absorption value of these two functional groups in PD-ST is significantly greater than that of PD-XG. The peaks at 1216 and 1270 cm^{-1} correspond to the stretching vibration peaks of the C-OH and CH_2OH groups, respectively, and the 1339, 1647, and 2932 cm^{-1} peaks are the CH and CH_2 groups [27]. The peak at 1733 cm^{-1} corresponds to the C=O group [27].

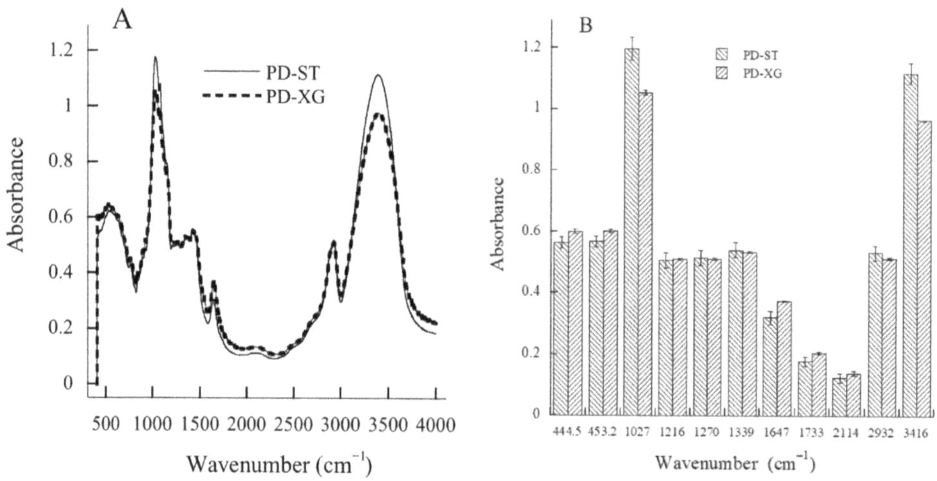

Figure 3. The absorbance of characteristic peaks in the FTIR of PD-ST and PD-XG polydextrose. Note: (**A**) IR scanning curves; (**B**) Characteristic peaks; Number of repetitions—n = 3.

Both types of polydextrose have similar purity levels and contents of small molecular sugars, complying with GB1986.385-2025 [28]. Compared with the moisture content (4.66 $\pm$ 0.03% wet basis) of PD-XG polydextrose, PD-ST shows a significant decrease in moisture content (1.00% $\pm$ 0.01%) and has high performance costs. The important process feature of PD-ST is that it was vacuum-dried during the entire polymerization reaction cycle in the reactor. Polydextrose is a versatile ingredient in the food industry, possibly due to its amorphous powder state (Figure 1) and strong water-holding capacity (Figure 2). The present study used PD-ST to carry out ana dditive test for wheat dough. The observed stronger hydroxyl peaks in PD-ST might be linked to increased hydrophilicity or a greater potential for hydrogen bonding with flour components.

2.2. Effect of Added Polydextrose on the Texture Profile and Appearance of Steamed Bread

Compared with the control sample, the 3–10% polydextrose PD-ST addition significantly increased the adhesiveness, hardness, gumminess, and chewiness of steamed bread, but the resilience, cohesiveness, and springiness basically remained the same, and the adhesive force was reduced (Table 1).

Table 1. Effect of adding polydextrose on the texture profiles of steamed bread.

PD-ST Addition (%)	Hardness (g)	Adhesive Force (g)	Adhesiveness (0.1 mJ)	Resilience (0.1)	Cohesiveness (0.1)	Springiness (mm)	Gumminess (g)	Chewiness (mJ)
0	1171 ± 78^{d}	3.05 ± 0.07^{a}	0.2 ± 0.0^{b}	4.13 ± 0.25^{ab}	7.80 ± 0.17^{ab}	10.84 ± 0.11^{a}	829 ± 35^{c}	87.7 ± 3.1^{c}
3	1067 ± 58^{d}	1.20 ± 1.13^{b}	0.2 ± 0.1^{b}	4.63 ± 0.42^{a}	8.03 ± 0.15^{a}	10.89 ± 0.25^{a}	872 ± 60^{b}	93.2 ± 7.7^{bc}
5	1394 ± 71^{a}	4.17 ± 3.18^{ab}	3.9 ± 0.3^{a}	3.97 ± 0.25^{ab}	7.77 ± 0.21^{ab}	10.90 ± 0.09^{a}	1081 ± 86^{a}	115.7 ± 10.1^{a}
7	1237 ± 84^{b}	1.00 ± 0.44^{b}	3.9 ± 0.2^{a}	3.97 ± 0.15^{b}	7.63 ± 0.06^{b}	10.78 ± 0.14^{a}	948 ± 59^{b}	100.3 ± 7.5^{b}
10	1244 ± 54^{c}	1.07 ± 0.15^{b}	4.7 ± 0.7^{a}	4.01 ± 0.21^{b}	7.61 ± 0.21^{b}	10.79 ± 0.11^{a}	1101 ± 64^{a}	117.8 ± 9.0^{a}

Note: All data are expressed as mean ± SD; number of repetitions—$n = 3$. Different superscript letters indicate the significant differences ($p < 0.05$) within the column.

The 3–10% polydextrose addition decreased the specific volume of steamed bread but increased its mass (Table 2). With respect to the steamed bread made from high-gluten wheat flour, the present study showed that polydextrose PD-ST increased their volumes at 7% and 10% addition and increased mass at 3–10% addition. These results are essentially similar to the results of Jing [10], where 1%, 5%, 7%, and 9% polydextrose additions significantly increased the volume and specific volume of steamed bread and 3%, 7%, and 9% polydextrose additions increased the mass of steamed bread (the steamed bread might be processed from medium-gluten flour). Our results suggest that, with an increase in polydextrose addition, raw steamed bread with the same weight will absorb more steam from the steamer during the steaming process, resulting in an increase in both the volume and weight of the steamed bread. The steamed bread containing 3–10% PD-ST had similar specific volumes, and this is possibly due to their similar volumes.

Table 2. Effect of added polydextrose on the specific volume of steamed bread.

PD-ST Addition (%)	Volume (cm^3)	Mass (g)	Specific Volume (cm^3/g)
0	190.1 ± 0.9^{b}	95.0 ± 0.5^{c}	2.00 ± 0.01^{a}
3	192.1 ± 1.2^{ab}	99.8 ± 0.9^{b}	1.92 ± 0.02^{b}
5	192.5 ± 2.5^{ab}	100.0 ± 1.3^{b}	1.92 ± 0.03^{b}
7	194.4 ± 2.1^{a}	102.5 ± 0.7^{a}	1.89 ± 0.02^{b}
10	195.5 ± 0.6^{a}	103.4 ± 0.5^{a}	1.89 ± 0.01^{b}

Note: All data are expressed as mean ± SD; number of repetitions—$n = 3$. Different superscript letters indicate significant differences ($p < 0.05$) within the column.

The brightness index, yellowish color, and color difference of steamed bread were significantly increased by the 5–10% polydextrose addition (Table 3). Compared with the control sample, 3%, 5%, 7%, and 10% polydextrose additions increased the color difference by 5.8%, 15.8%, 14.6%, and 16.0%, respectively.

Table 3. Effect of adding polydextrose on the color of steamed bread.

PD-ST (%)	L	−a	b	ΔE
0	77.39 ± 0.54 ^{b}	1.707 ± 0.029 ^{a}	12.89 ± 0.25 ^{b}	21.03 ± 0.66 ^{c}
3	78.40 ± 0.39 ^{b}	1.642 ± 0.023 ^{b}	13.32 ± 0.17 ^{a}	22.34 ± 0.39 ^{b}
5	80.44 ± 0.51 ^{a}	1.502 ± 0.046 ^{c}	13.44 ± 0.28 ^{a}	24.35 ± 0.51 ^{a}
7	80.15 ± 0.31 ^{a}	1.697 ± 0.023 ^{a}	13.51 ± 0.12 ^{a}	24.09 ± 0.30 ^{a}
10	80.33 ± 0.12 ^{a}	1.721 ± 0.011 ^{a}	13.55 ± 0.11 ^{a}	24.39 ± 0.28 ^{a}

Note: L indicates the brightness index, a indicates greenish tones, b indicates yellowish tones, and ΔE indicates a color difference. All data are expressed as mean ± SD; number of repetitions—n = 3. Different superscript letters indicate significant differences ($p < 0.05$) within the column.

2.3. *Effect of Added Polydextrose on the Sensory Attributes of Steamed Bread*

Table 4 shows that the 3–10% polydextrose addition significantly reduced the specific volume and width/height ratio of the steamed bread; improved its internal structure; and maintained the elasticity, surface color and structure, toughness, adhesiveness, taste, and total score. The panel for the sensory analysis was trained by the experienced experts. The cross-sections in Figure 4 show that the steamed bread became uniform in texture, with fewer holes appearing as the amount of polydextrose added increased.

Table 4. Effect of added polydextrose on the sensory attributes of steamed bread.

PD-ST (%)	Specific Volume	Width/Height Ratio	Elasticity	Surface Color	Surface Structure
0	8.00 ± 0.00 ^{a}	4.00 ± 0.00 ^{a}	7.00 ± 0.58 ^{a}	7.00 ± 0.58 ^{a}	7.40 ± 0.73 ^{a}
3	7.00 ± 0.00 ^{b}	3.00 ± 0.00 ^{b}	7.60 ± 0.45 ^{a}	6.80 ± 0.68 ^{a}	7.80 ± 0.37 ^{a}
5	7.00 ± 0.00 ^{b}	3.00 ± 0.00 ^{b}	7.20 ± 1.06 ^{a}	6.80 ± 0.37 ^{a}	8.40 ± 0.45 ^{a}
7	6.50 ± 0.00 ^{c}	3.00 ± 0.00 ^{b}	7.40 ± 0.45 ^{a}	6.80 ± 0.89 ^{a}	7.60 ± 0.93 ^{a}
10	6.50 ± 0.00 ^{c}	3.00 ± 0.00 ^{b}	6.80 ± 0.37 ^{a}	7.40 ± 0.73 ^{a}	7.40 ± 0.73 ^{a}
PD-ST (%)	**Internal Structure**	**Toughness**	**Adhesiveness**	**Taste**	**Total Score**
0	15.40 ± 0.45 ^{b}	7.40 ± 0.93 ^{a}	7.60 ± 0.93 ^{a}	4.60 ± 0.45 ^{a}	68.40 ± 2.05 ^{a}
3	16.20 ± 0.37 ^{ab}	6.80 ± 0.68 ^{a}	7.80 ± 0.68 ^{a}	4.60 ± 0.45 ^{a}	67.60 ± 2.21 ^{a}
5	16.20 ± 1.57 ^{ab}	7.40 ± 0.45 ^{a}	7.60 ± 0.45 ^{a}	5.00 ± 0.00 ^{a}	68.60 ± 1.59 ^{a}
7	16.40 ± 1.48 ^{ab}	7.00 ± 0.82 ^{a}	8.20 ± 0.37 ^{a}	4.80 ± 0.37 ^{a}	67.70 ± 2.61 ^{a}
10	17.20 ± 1.21 ^{a}	7.60 ± 0.93 ^{a}	8.40 ± 0.73 ^{a}	4.80 ± 0.37 ^{a}	69.10 ± 1.69 ^{a}

Note: All data are expressed as mean ± SD; number of repetitions—n = 8. Different superscript letters indicate the significant differences ($p < 0.05$) within the column.

2.4. *Effect of Added Polydextrose on the Pasting and Thermal Parameters of Wheat Flours*

Table 5 shows that polydextrose additions decreased the peak, trough, and final viscosity, as well as the breakdown and setback viscosity of the pasting of wheat flour suspension solutions, but increased the pasting temperature; the gelatinization peak time tended to reduce.

Table 5. Effect of adding polydextrose on the pasting parameters of wheat flours.

PD-ST (%)	Peak Viscosity (cp)	Trough Viscosity (cp)	Breakdown Viscosity (cp)	Final Viscosity (cp)	Setback Viscosity (cp)	Peak Time (min)	Pasting Temp. (°C)
0	2061 ± 15 ^{a}	1228 ± 7 ^{a}	833 ± 18 a	2522 ± 8 ^{a}	1293 ± 14 ^{a}	6.11 ± 0.03 ^{a}	69.33 ± 0.08 ^{d}
3	1854 ± 8 ^{b}	1125 ± 15 ^{b}	729 ± 10 ^{b}	2322 ± 10 ^{b}	1197 ± 23 ^{b}	6.09 ± 0.08 ^{ab}	69.83 ± 0.51 ^{cd}
5	1806 ± 6 ^{c}	1095 ± 13 ^{c}	711 ± 19 ^{b}	2285 ± 3 ^{c}	1190 ± 13 ^{b}	6.09 ± 0.03 ^{a}	70.13 ± 0.03 ^{c}
7	1634 ± 17 ^{d}	1000 ± 28 ^{d}	634 ± 18 ^{c}	2096 ± 6 ^{d}	1096 ± 31 ^{c}	6.04 ± 0.10 ^{ab}	87.42 ± 0.43 ^{b}
10	1481 ± 13 ^{e}	913 ± 16 ^{e}	569 ± 4 ^{d}	1943 ± 10 ^{e}	1030 ± 9 ^{d}	6.00 ± 0.01 ^{b}	87.95 ± 0.09 ^{a}

Note: All data are expressed as mean ± SD; number of repetitions—n = 3. Different superscript letters indicate significant differences ($p < 0.05$) within the column.

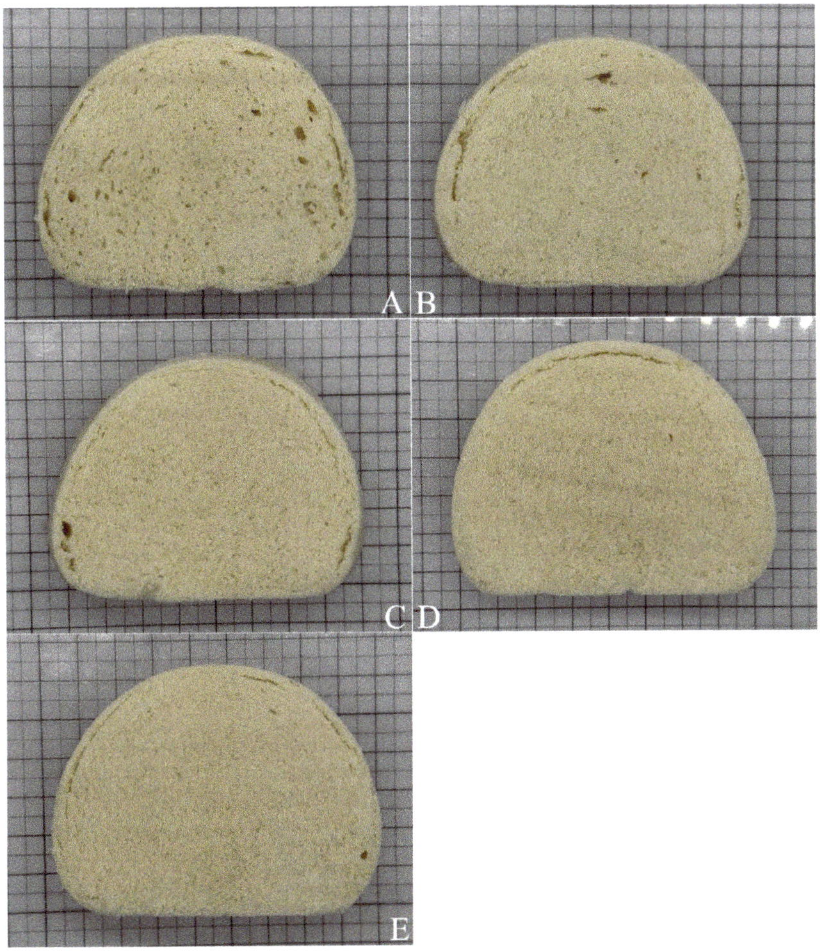

Figure 4. Cross-section of steamed bread. Note: (**A**) is the control without polydextrose; (**B–E**) are 3%, 5%, 7%, and 10% PD additions. All photos are enlarged 1×.

Table 6 shows that when determining the thermal properties of flour paste at 0 d, compared with the control sample, the addition of polydextrose significantly reduced the enthalpy value and peak width of gelatinization but increased the onset temperature, peak temperature, and conclusion temperature of gelatinization; the peak width of gelatinization essentially remained the same. The PD-ST addition increased the peak temperature of gelatinization in wheat flours, which is similar to the results of Rosenthal [29], where polydextrose elevated the starch gelatinization temperature. Our study suggests that polydextrose, with its high water-binding capacity, could compete with starch for water molecules during gelatinization, leading to reduced peak viscosity and gelatinization enthalpy.

Table 7 further shows that when the retrograded samples were kept at 4 °C for 21 days and the thermal properties were then re-measured, compared with the control sample, the addition of polydextrose significantly reduced the enthalpy value, peak width, and peak width of gelatinization but increased the onset temperature; the peak temperature of gelatinization remained unchanged. The aging rate of wheat flour paste was significantly reduced by the addition of polydextrose.

Table 6. Effect of adding polydextrose on the thermal properties of wheat flour (0 d).

PD-ST (%)	ΔH (J/g)	T_p (°C)	T_o (°C)	T_c (°C)	Peak Width (°C)	Peak Height (0.01 mW/mg)
0	5.57 ± 0.25 [a]	63.27 ± 0.15 [d]	57.47 ± 0.06 [c]	70.00 ± 0.40 [bc]	5.90 ± 0.10 [a]	11.77 ± 0.80 [ab]
3	5.37 ± 0.02 [a]	63.60 ± 0.17 [c]	57.70 ± 0.20 [bc]	70.00 ± 0.20 [c]	5.80 ± 0.10 [ab]	11.77 ± 0.30 [a]
5	5.10 ± 0.11 [b]	63.97 ± 0.25 [bc]	57.93 ± 0.15 [b]	70.30 ± 0.26 [bc]	5.80 ± 0.01 [ab]	11.36 ± 0.18 [a]
7	4.88 ± 0.11 [bc]	64.67 ± 0.68 [ab]	58.73 ± 0.21 [a]	71.00 ± 0.66 [ab]	5.87 ± 0.15 [ab]	10.68 ± 0.34 [b]
10	4.81 ± 0.11 [c]	64.57 ± 0.15 [a]	58.77 ± 0.06 [a]	70.67 ± 0.06 [a]	5.73 ± 0.06 [b]	10.84 ± 0.30 [b]

Note: ΔH, Gelatinization enthalpy; T_o, T_p, and T_c are the onset temperature, peak temperature, and conclusion temperature of gelatinization, respectively. All data are expressed as mean ± SD; number of repetitions—n = 3. Different superscript letters indicate significant differences ($p < 0.05$) within the column.

Table 7. Effect of adding polydextrose on the thermal properties of retrograded samples (21 d).

PD-ST (%)	ΔH (J/g)	T_p (°C)	T_o (°C)	T_c (°C)	Peak Width (°C)	Peak Height (0.01 mW/mg)	Aging (%)
0	1.43 ± 0.19 [a]	53.80 ± 0.85 [ab]	46.57 ± 0.21 [c]	62.53 ± 0.21 [b]	8.80 ± 0.53 [a]	2.30 ± 0.28 [a]	25.72 ± 4.35 [a]
3	1.05 ± 0.13 [b]	53.20 ± 3.40 [ab]	48.20 ± 0.01 [b]	62.47 ± 0.25 [bc]	8.50 ± 0.20 [a]	1.87 ± 0.12 [b]	19.58 ± 2.51 [ab]
5	0.82 ± 0.09 [b]	52.60 ± 0.61 [b]	49.30 ± 0.10 [a]	63.73 ± 0.25 [a]	7.80 ± 0.26 [b]	1.59 ± 0.14 [c]	16.19 ± 2.15 [b]
7	0.82 ± 0.01 [c]	53.83 ± 0.71 [ab]	48.27 ± 0.80 [ab]	61.97 ± 0.25 [c]	7.90 ± 0.36 [b]	1.54 ± 0.06 [c]	16.80 ± 0.40 [b]
10	0.77 ± 0.04 [c]	54.50 ± 0.01 [a]	49.23 ± 0.98 [a]	62.03 ± 0.25 [c]	7.60 ± 0.46 [b]	1.52 ± 0.08 [c]	16.11 ± 1.10 [b]

Note: ΔH, Gelatinization enthalpy; T_o, T_p, and T_c are the onset temperature, peak temperature, and conclusion temperature of gelatinization, respectively. All data are expressed as mean ± SD; number of repetitions—n = 3. Different superscript letters indicate significant differences ($p < 0.05$) within the column.

2.5. Effect of Added Polydextrose on the Thermo-Mechanical Property of Wheat Dough Without Yeast Powder

Table 8 shows that with an increase in the amount of added polydextrose, the dough's development time significantly increased, and the dough stability time first increased and then decreased, reaching a maximum at 5% polydextrose addition. The protein's weakening degree, peak torque of gelatinization, starch breakdown, and starch setback torque all increased while α-amylase and enzymatic hydrolysis rates remained unchanged. The heating rate increased, and the gelatinization rate first increased and then decreased. The 3–10% PD-ST addition decreased the dough's protein weakness and starch gelatinization torque, but the dough's gelatinization speed was retained or even increased by the 3–7% PD-ST addition, and it decreased via the10% PD-ST addition, suggesting that PD-ST additions could reinforce the gluten matrix, increase pasting temperatures, and decrease the viscosity of flour through hydration effects.

2.6. Effect of Added Polydextrose on Starch Crystallinity and Protein Conformation

FTIR analysis showed that the $R_{1022/995}$ of unfermented dough was significantly greater than that of the steamed bread, but its $R_{1047/1022}$ and $R_{1068/1022}$ ratios were significantly lower than those of steamed bread (Tables 9 and 10). Moreover, the percentage of β-sheets in unfermented dough was significantly lower than that in steamed bread, but its contents of random coils, α-helixes, and β-turns were significantly higher. These results suggest that, in the presence of a proper amount of water molecules, the higher amorphous regions of starch in the dough, as well as the higher relative contents of secondary structures like random coils, α-helixes, and β-turns, resulted in the orderly state of the gluten network structure.

Table 8. Effect of added polydextrose on the thermo-mechanical parameters of wheat dough without yeast powder.

PD-ST (%)	DDT (min)	DST (min)	C1–Cs (0.1 Nm)	C3 (Nm)	C3/C4
0	3.55 ± 0.36 [e]	4.83 ± 0.42 [d]	2.28 ± 0.17 [a]	1.57 ± 0.02 [a]	1.149 ± 0.011 [ab]
3	4.55 ± 0.17 [d]	7.03 ± 0.32 [c]	0.74 ± 0.09 [b]	1.55 ± 0.02 [a]	1.151 ± 0.014 [ab]
5	5.32 ± 0.34 [c]	8.30 ± 0.10 [a]	0.27 ± 0.07 [c]	1.46 ± 0.01 [b]	1.126 ± 0.006 [c]
7	6.91 ± 1.05 [b]	7.27 ± 0.78 [b]	0.05 ± 0.04 [d]	1.46 ± 0.01 [b]	1.126 ± 0.014 [bc]
10	8.59 ± 0.06 [a]	5.25 ± 0.07 [d]	0.08 ± 0.02 [d]	1.32 ± 0.01 [c]	1.159 ± 0.009 [a]
PD-ST (%)	**C3–C4 (0.1 Nm)**	**C5–C4 (0.1 Nm)**	**α (−0.01 Nm)**	**β (0.01 Nm)**	**γ (−0.01 Nm)**
0	2.05 ± 0.12 [a]	6.74 ± 0.10 [a]	4.93 ± 0.31 [c]	46.73 ± 1.15 [b]	2.80 ± 1.91 [a]
3	2.03 ± 0.16 [ab]	7.05 ± 0.43 [a]	4.73 ± 0.81 [bc]	45.27 ± 1.80 [b]	3.07 ± 1.15 [a]
5	1.63 ± 0.08 [d]	6.23 ± 0.15 [b]	5.47 ± 0.81 [bc]	45.93 ± 4.69 [abc]	2.33 ± 1.67 [a]
7	1.62 ± 0.16 [cd]	6.04 ± 0.13 [b]	6.00 ± 0.72 [ab]	50.47 ± 0.61 [a]	14.87 ± 1.96 [a]
10	1.81 ± 0.08 [bc]	5.11 ± 0.08 [c]	6.50 ± 0.14 [a]	41.70 ± 0.99 [c]	3.10 ± 0.14 [a]

Note: DDT, Dough development time; DST, dough stability time; C1–Cs, protein weakness; C3, maximum gelatinization torque; C3–C4, starch breakdown; C3/C4, amylase activity; C5–C4, starch setback; α, heating speed; β, gelatinization speed; γ, enzymatic degradation speed. All data are expressed as mean ± SD; number of repetitions—$n = 3$. Different superscript letters indicate significant differences ($p < 0.05$) within the column.

Table 9. Effect of polydextrose on starch crystallinity and protein conformation in dough.

PD-ST (%)	$R_{1022/995}$	$R_{1047/1022}$	$R_{1068/1022}$	
0	1.0419 ± 0.0006 [b]	0.9147 ± 0.0022 [d]	0.7492 ± 0.0038 [e]	
3	1.0334 ± 0.0004 [c]	0.9259 ± 0.0019 [a]	0.7840 ± 0.0049 [a]	
5	1.0333 ± 0.0015 [c]	0.9227 ± 0.0007 [b]	0.7742 ± 0.0023 [b]	
7	1.0426 ± 0.0004 [b]	0.9185 ± 0.0009 [c]	0.7628 ± 0.0012 [d]	
10	1.0486 ± 0.0017 [a]	0.9189 ± 0.0002 [c]	0.7688 ± 0.0001 [c]	
PD-ST (%)	**β-sheet (%)**	**Random coil (%)**	**α-helix (%)**	**β-turn (%)**
0	48.928 ± 0.102 [d]	17.291 ± 0.043 [a]	17.281 ± 0.025 [a]	16.499 ± 0.035 [a]
3	50.244 ± 0.258 [b]	16.928 ± 0.073 [d]	16.807 ± 0.088 [c]	16.022 ± 0.096 [c]
5	49.507 ± 0.146 [c]	17.141 ± 0.041 [c]	17.073 ± 0.061 [b]	16.278 ± 0.045 [b]
7	49.361 ± 0.065 [c]	17.202 ± 0.014 [b]	17.145 ± 0.031 [b]	16.292 ± 0.021 [b]
10	50.184 ± 0.037 [a]	16.992 ± 0.015 [d]	16.826 ± 0.007 [c]	15.998 ± 0.020 [c]

Note: PD, Polydextrose; $R_{1022/995}$ and $R_{1047/1022}$ show the degree of short sequences at the surface of starch granules, and $R_{1068/1022}$ shows the interaction between starch and proteins. All data are expressed as mean ± SD; number of repetitions—$n = 3$. Different superscript letters indicate significant differences ($p < 0.05$) within the column.

Table 10. Effect of polydextrose on starch crystallinity and protein conformation in steamed bread.

PD-ST (%)	$R_{1022/995}$	$R_{1047/1022}$	$R_{1068/1022}$	
0	1.0249 ± 0.0005 [a]	0.9402 ± 0.0037 [c]	0.8316 ± 0.0083 [c]	
3	1.0183 ± 0.0010 [bc]	0.9473 ± 0.0018 [b]	0.8554 ± 0.0017 [b]	
5	1.0175 ± 0.0008 [c]	0.9501 ± 0.0034 [b]	0.8644 ± 0.0076 [b]	
7	1.0201 ± 0.0010 [b]	0.9351 ± 0.0024 [c]	0.8306 ± 0.0011 [c]	
10	1.0162 ± 0.0016 [c]	0.9565 ± 0.0021 [a]	0.8784 ± 0.0045 [a]	
PD-ST (%)	**β-sheet (%)**	**Random coil (%)**	**α-helix (%)**	**β-turn (%)**
0	52.587 ± 0.348 [bc]	16.345 ± 0.081 [b]	15.853 ± 0.126 [b]	15.214 ± 0.142 [bc]
3	52.933 ± 0.235 [b]	16.283 ± 0.081 [b]	15.605 ± 0.134 [bc]	15.178 ± 0.123 [c]
5	52.456 ± 0.146 [c]	16.267 ± 0.026 [b]	15.886 ± 0.179 [b]	15.391 ± 0.072 [b]
7	52.669 ± 0.106 [bc]	16.467 ± 0.014 [a]	16.169 ± 0.057 [a]	15.694 ± 0.035 [a]
10	53.599 ± 0.202 [a]	16.074 ± 0.064 [c]	15.489 ± 0.060 [c]	14.837 ± 0.079 [d]

Notes: PD, Polydextrose; $R_{1022/995}$ and $R_{1047/1022}$ show the degree of short sequences at the surface of starch granules, and $R_{1068/1022}$ shows the interaction between starch and proteins. All data are expressed as mean ± SD; number of repetitions—$n = 3$. Different superscript letters indicate significant differences ($p < 0.05$) within the column.

Compared with the control sample, the addition of polydextrose increased the amorphous and crystalline regions of starch and the interaction between proteins and starch in wheat dough without yeast (Table 9). The percentage of β-sheets accounted for around 50% of the secondary protein structure in the wheat dough. The 3–7% polydextrose addition significantly increased the β-sheets but decreased the random coils, α-helixes, and β-turns.

For steamed bread, the addition of polydextrose decreased the amorphous regions of starch but increased the crystalline regions of starch and the interaction between protein and starch (Table 10). Compared with the dough, the β-sheet conformation in steamed bread was increased by 2%, but random coils, α-helixes, and β-turns were reduced. The 3–7% polydextrose addition significantly increased the amount of β-sheets and the percentages of random coils, α-helixes, and β-turns in steamed bread, but the 10% polydextrose addition decreased the conformations of random coils, α-helixes, and β-turns. These results of moderate PD-ST (3–7%) could promote ordered secondary protein structures, while high PD-ST (10%) might disrupt protein folding, possibly due to excessive water competition or steric hindrance.

2.7. *Effect of Added Polydextrose on the Microstructure of Unfermented Dough*

When the freeze-dried cubes of unfermented dough were observed at an enlargement of 100 times, the surface of the dough was uneven, and it occasionally had larger and deeper holes with irregular openings (Figure 5A). With an increase in the addition concentrations of polydextrose, the holes on the surface of the dough increased significantly (Figure 5B,C), the openings were approximately circular, and the holes gradually deepened (Figure 5D). In particular, with the 10% PD-ST addition, the number of holes increased, holes with larger openings increased, and the holes became deeper (Figure 5E). These results are basically similar to the results shown in Figure 4. These results suggest that the hydroxyl groups in strongly hydrophilic polydextrose molecules might combine with the carboxyl groups of gluten proteins, resulting in an even distribution of polydextrose molecules in the gluten network. When the samples for electron microscope observation were prepared by the freeze-drying method, the water molecules escaped, leaving holes with relatively uniform diameters in the gluten network. These conclusions can be further confirmed in wheat flour dough using laser confocal scanning microscopy (CLSM) or quantified with image analysis software.

Figure 6 further shows the microstructure of the dough at an enlargement of 1200 times. In the dough without added polydextrose, spherical or ellipsoidal starch granules were exposed on the surface, in the interstices of the solidified gluten protein network, or embedded in the gluten protein network, with clear gaps between the starch granules and clear boundaries (Figure 6A). Polydextrose relies on its unique structure to form a continuous three-dimensional network structure with certain adhesion and elasticity, with the 3–7% polydextrose addition playing a similar role to the gluten network structure, gradually increasing the uniform pores and even the bubble-holding capacity of the dough (Figure 6B–D). At 10% polydextrose addition, the adhesion, freshness, and looseness of the dough were further increased, the holes became smaller and more numerous, and the boundaries between starch granules became blurred (Figure 6E).

Dietary fiber is more widely used in baked foods in Western countries [7]. The amount of added dietary fiber is generally 5–10% of the flour content. If the amount exceeds 10%, it will slow down the fermentation speed of the bread [30,31]. In the present study, the 3–10% polydextrose (PD-ST) addition was tested in steamed breads made from high-gluten wheat flours. These addition levels maintained most of the sensory evaluation parameters of the steamed bread, such as its elasticity, surface color and structure, toughness, adhesiveness, taste, and total score, despite significantly reducing the specific volume and width/height ratio of the bread. These results are similar to the results of Jing et al. [10], where the 1–9% polydextrose addition had little effect on the sensory evaluation of steamed bread (their bread may be made from middle-gluten wheat flours). Our result is different from that of a previous study where the addition of konjac glucomannan was found to reduce the hardness of steamed bread [32], possibly due to the hygroscopic properties of different polysaccharides [33]. The increase in the brightness, color difference, hardness, adhesiveness, gumminess, and chewiness of steamed bread by the addition of 5–10% polydextrose could be explained by the three-dimensional network structure of the unfermented dough with certain adhesion and freshness properties, as observed under SEM.

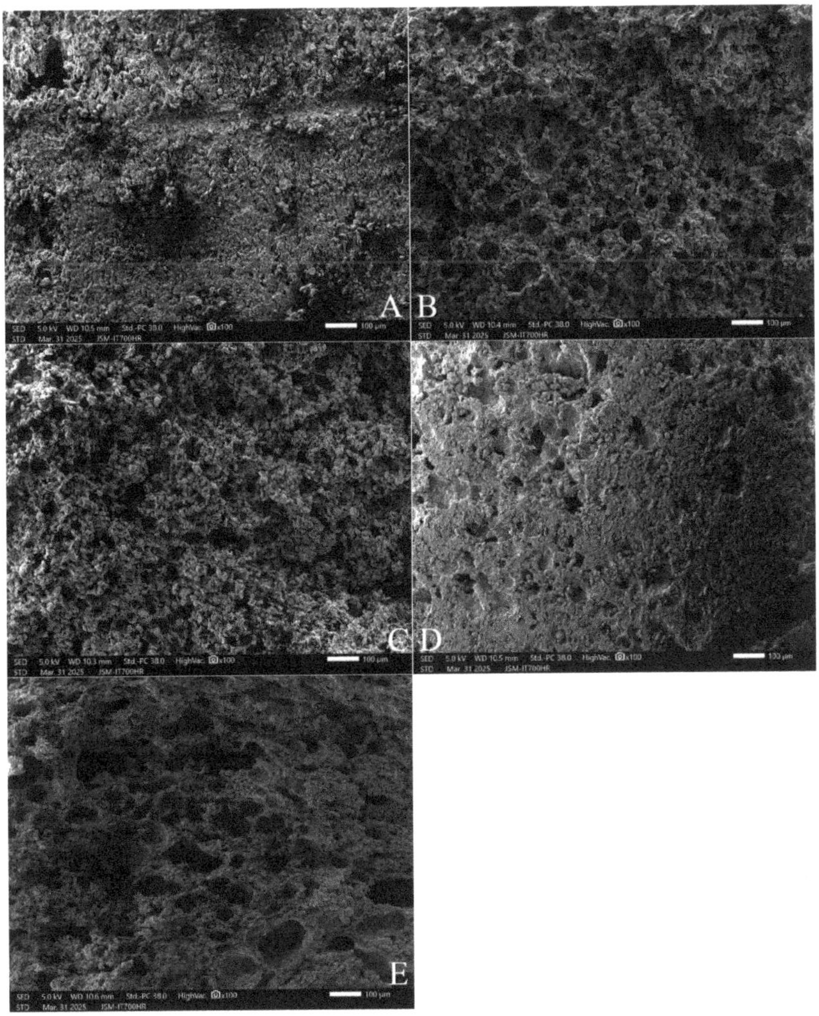

Figure 5. Microstructure of the freeze-dried cubes of unfermented dough (100×).Note: (**A**) is the control without polydextrose; (**B–E**) are 3%, 5%, 7%, and 10% PD-ST additions. All photos are enlarged 100×.

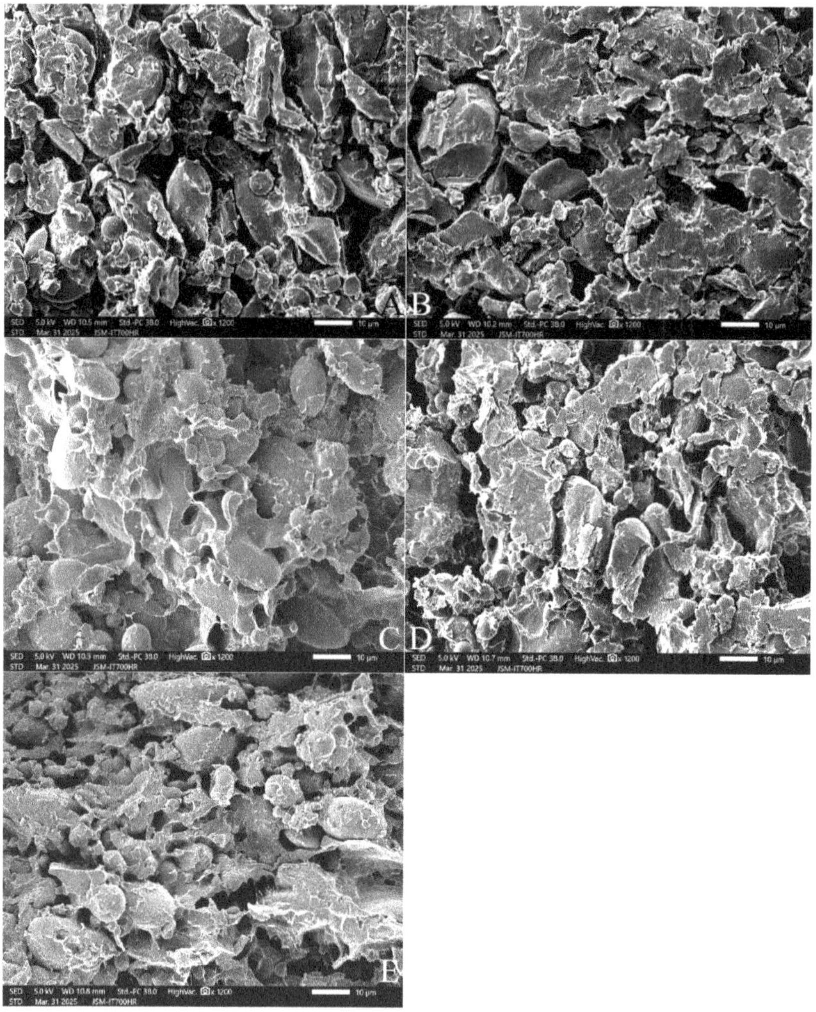

Figure 6. Microstructure of the freeze-dried cubes of unfermented dough (1200×).Note: (**A**) is the control without polydextrose; (**B–E**) are 3%, 5%, 7%, and 10% PD-ST additions. All photos are enlarged to 1200×.

Polydextrose has been incorporated into a wide range of foods, including baked goods, beverages, confectionery, and frozen desserts [17]; however, few studies have meticulously investigated the effects of polydextrose addition in steamed bread. Among 1–9% polydextrose additions, Jing et al. [10] reported that a 3% polydextrose addition improved the maximal hardness and chewiness, while other addition amounts resulted in no changes relative to the control sample. In the present study, the 3–10% polydextrose addition to steamed bread resulted in similar textural parameters, like springiness, cohesiveness, and resilience, to the control sample, except for the increase in hardness, adhesiveness, gumminess, and chewiness and a decreasing trend in adhesive force. Schirmer et al. [18] added 0–21.96% polydextrose into the batter of pound cake as a replacement for sucrose and found that the firmness of the batter was unchanged, but its adhesive force significantly increased. Further, the firmness and springiness of the pound cake remained unchanged with 0–17.57% polydextrose additions, but they were significantly reduced with a 21.96%

polydextrose addition. Additionally, 0–21.96% polydextrose additions significantly increased the specific volume of the cake. These results suggest that 0–10% polydextrose additions to steamed bread or baked pound cake retain the cohesiveness and springiness of the product. Further studies should explore the effects of 20% polydextrose additions on the textural parameters of steamed bread.

Our recent study showed that the 3–10% polydextrose addition significantly decreased the hardness, adhesive force, adhesiveness, cohesiveness, gumminess, and chewiness of cooked Indica rice [34]. In the present study, the 3–10% polydextrose addition to steamed bread resulted in an increase in hardness, adhesiveness, gumminess and chewiness but retained similar textural parameters, like springiness, cohesiveness, and resilience, as the control sample, suggesting that polydextrose might play different roles in the interaction between starch and protein in cooked rice versus steamed bread through intermolecular interactions and the formation of hydrogel-like networks.

Early analyses of the microstructure of cake batters with polydextrose substitutions showed that the addition of 14% and 28.1% polydextrose in cake batters resulted in a better air-holding capacity, which resulted in an increase in bubble size [35], but Hicsasmaz et al. [36] showed that an increase in the polydextrose content from 0% to 28.1% led to a decrease in the number and size of crack-like pores, with an increase in bubble count in the cake batter. Kocer et al. [37] added 0–28.1% polydextrose to replace sugar in a high-ratio cake and found that the uniformity of the pore sizes in the crumb increased with increasing amounts of added polydextrose. In the present study, a 10% polydextrose addition resulted in significantly more sphere-like pores than crack-like pores in the unfermented dough cube, along with increased uniformity of pore sizes. The adhesion, freshness, and looseness of the dough increased with the amount of polydextrose added. We conclude that with an increase in polydextrose addition, raw steamed bread of the same weight will absorb more steam from a steamer during the steaming process, resulting in an increase in both the volume and weight of steamed bread, as well as improved freshness and looseness.

In the present study, the relative percentage of β-sheets in unfermented dough was significantly lower than that in steamed bread, but the α-helix and β-turn percentages were significantly higher than those in steamed bread, which is consistent with the report of Ding et al. [38].The content of random coils in the unfermented dough in our study was higher than that in the steamed bread. This result is different from that of Ding et al. [38], possibly because we used infrared spectroscopy in the range of 1670–1600 cm^{-1} to analyze the secondary structure of protein, while Ding et al. [38] used infrared spectroscopy in the range of 1330–1220 cm^{-1} to analyze the secondary structure of proteins [39].

Polydextrose additions can prevent baked products from spoiling and extend their shelf life [22]. In the present study, the 3–10% polydextrose addition significantly decreased the setback viscosity of high-gluten wheat flour and the aging of retrograded flour paste, and 5–10% polydextrose reduced the starch setback torque (C5–C4) of dough, suggesting that polydextrose additions can prevent steamed bread from aging during storage and maintain its springiness and cohesiveness. Further studies are needed to define the effect of polydextrose (PD-ST) on the textural profile of steamed bread made from middle- or low-gluten wheat flours, as well as on the dough microstructure, pasting, thermal, and thermal mechanical properties, and to determine the best amount of polydextrose addition.

3. Conclusions

The addition of 3–10% polydextrose (PD-ST) to high-gluten wheat flour modified its rheological, thermal, and thermo-mechanical properties. It decreased the peak, breakdown, and setback viscosity of pasting; reduced the enthalpy of gelatinization and the aging rate of flour paste while increasing the peak temperature of gelatinization; and decreased the protein weakening degree, peak torque of gelatinization, starch breakdown, and starch setback torque while increasing dough development time.

The addition of polydextrose increased the crystalline regions of starch, the interaction between protein and starch, and the β-sheet percentage of wheat dough and steamed bread. The addition of polydextrose also increased the amorphous regions of starch in the dough but decreased these in steamed bread. Further, the 3–10% polydextrose addition decreased the percentages of random coils, α-helixes, and β-turns in the dough, but the 3–7% polydextrose addition maintained or increased these conformations in steamed bread, while 10% polydextrose decreased them.

The 3–10% polydextrose addition decreased the specific volume and width/height ratio of the steamed bread but increased the brightness index, hardness, adhesiveness, gumminess, and chewiness. The microstructure results indicated that the 3–10% polydextrose addition resulted in the formation of a continuous three-dimensional network structure with certain adhesion and elasticity in the unfermented dough, increasing the number of small pores and the gas-holding capacity of the product through unique molecular interactions. Polydextrose could alter dough rheology and the texture of steamed bread by competing with starch for water and interfering with gluten formation.

In summary, the addition of 3–10% polydextrose to steamed bread not only increased the dietary fiber content of the product but also improved its chewiness and shelf life.

4. Materials and Methods

4.1. Samples

Polydextrose ST (PD-ST) with an initial moisture content of 1.0% was provided by Runloy Biotechnology (Shanghai) Co., Ltd., Shanghai, China. Polydextrose XG (PD-XG) was provided by Xingguang Co., Ltd., Binzhou, China. High-gluten wheat core flour (Golden Dragonfish, Wilmar Shanghai Co., Ltd., Shanghai, China) was purchased from a local supermarket. PD-ST was added to wheat flour at mass ratios of 0%, 3%, 5%, 7%, and 10%, respectively, to obtain polydextrose-added samples. The measured analysis parameters and important experimental steps are shown in Figure 7.

4.2. The Moisture Absorption Rate of Polydextrose

The method for determining the moisture absorption of the sample is as follows: 1 g of polydextrose was weighed in an aluminum box (diameter of 5 cm and height of 3 cm) and placed on a steel wire rack in the middle of an incubator (26 °C). A ceramic plate filled with distilled water was placed at the bottom of the insulated oven incubator (DHG 9070A, Blue Sky Chemical Instrument Factory, Hangzhou, China), and the weight of the aluminum box containing the sample was weighed every half an hour for 12 h. The relative humidity was 84.1%, which was measured by a hygrometer.

4.3. Textural Profile of Wheat Steamed Bread

The textural profile of wheat steamed bread was analyzed with a textural analyzer (CTX, Brookfield, Middleboro, MA, USA), as described by Wang et al. [34]. Wheat steamed bread was made according to GB/T 35991-2018 [40]. A flour sample (200 g) was weighed and mixed with 40 g of warm water (38 °C) containing 1.6 g of yeast. The dough was then kneaded for 180 s in a kneading machine at room temperature after the addition of 50 g of deionized water. Subsequently, the dough was taken out and pressed into films 6–8 times using a rolling machine to move the bubbles. It was then divided into three equal parts. Subsequently, the dough was kneaded and folded towards the bottom edge to ensure that the front and back of the dough were kneaded the same number of times. The well-kneaded dough was fermented for 30 min at 35 °C and 85% relative humidity in a fermentation box. Then, 1500 mL of deionized water was added toa pot, and the bread was steamed for 25 min at 1600 W. It was then braised for 5 min without uncovering the pot's lid. The steamed bread was then removed and covered with gauze to cool at room temperature for 1 h before the later determination of textural profiles.

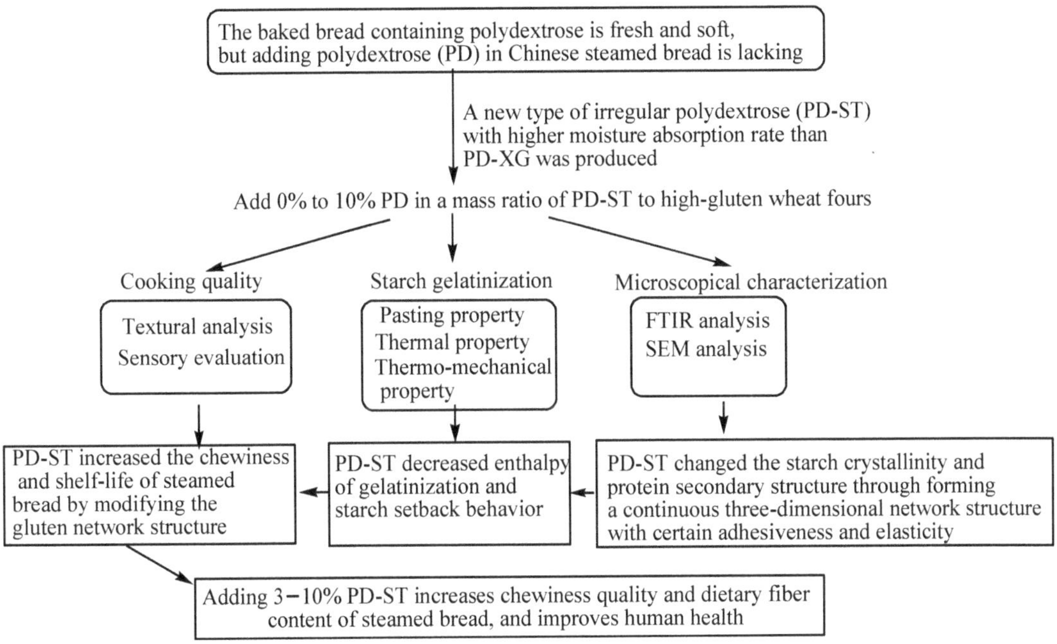

Figure 7. A schematic diagram of the present study.

4.4. Specific Volume and Color of Steamed Bread

The specific volume of steamed bread was determined according to GB/T 21118-2007 [41].The mass of the steamed bread was measured after cooling for 1 h. The volume of the steamed bread was measured by the rapeseed replacement method; the average value was taken after three repetitions. The specific volume was calculated according to the following formula:

$$\text{Specific volume}\left(\frac{\text{cm}^3}{\text{g}}\right) = \frac{\text{Volume of a steamed bread }(\text{cm}^3)}{\text{Mass of a steamed bread }(\text{g})} \quad (1)$$

The color of steamed bread was determined using a colorimeter (CR-100, Japanese Konica Minolta, Ltd., Shanghai, China). The steamed bread was cut into slices with a thickness of 15 mm using a slicer to obtain test samples. A standard white plate was used for calibration. Then, steamed bread slices were placed on the test port, completely covering it, and the test key was pressed. The sample was removed after the test was completed to obtain a set of color differences. The colorimeter measured the brightness (L*), redness-greenness (a*), and yellowness-blueness (b*) values using the color difference between the standard white plate and the tested steamed bread slices. ΔE is the overall color difference.

The slices of steamed bread were photographed using a mobile phone (Apple, iPhone 15, 12 million pixels, Apple official flagship store, Beijing, China).

4.5. Pasting Characteristics

An RVA—TecMaster (PertenRuihua Scientific Instrument Co., Ltd., Beijing, China) was used to determine the pasting parameters of the samples according to the GB/T24852-2010 method [42]. During measurement, the stirring paddle speed was set at 960 r/min for the initial 10 s and then lowered to 160 r/min within 20 s and kept at 180 r/min. The temperature was initially set at 50 °C for 1 min and then raised to 95 °C within 3.71 min and kept there for 2.5 min before being reduced to 50 °C within 2.8 min and retained for 2 min.

4.6. Thermal Properties

Polydextrose was uniformly mixed with wheat flour on amass basis to achieve addition levels of 0% to 10%. A 5.0 mg sample was employed to determine the gelatinization parameters using a differential scanning calorimeter (DSC 214, NetzschGmbH, Selb, Germany), according to the method of Wang et al. [34]. After gelatinization, the retrograded samples were placed in 24-well plates with covers at 4 °C for 21 days and again measured for retrogradation. The results were evaluated using the same analysis software. The aging of the retrograded wheat flour paste was calculated using Equation (2):

$$\text{Ageing (\%)} = (\text{Gelatinization enthalpy measured at day 21})/(\text{Gelatinization enthalpy measured at day 0}) \times 100 \quad (2)$$

4.7. Thermo-Mechanical Characteristics

A Mixolab (Chopin Technologies, Tripette et Renaud, Paris, France) was employed to analyze the thermo-mechanical characteristics of the dough, as described by Wang et al. [34]. For assays at constant hydration, about 50 g of wheat flour was placed into the Mixolab bowl. Then, 75 g of dough with 60% water content (14% moisture basis) was evaluated.

4.8. Fourier Transform Infrared Spectroscopy (FTIR)

The FTIR spectra of the cooked samples were recorded on a Nicolet 6700 FTIR spectrometer (Thermo Fisher Scientific, Greater Mumbai, MA, USA). The test conditions were as follows: spectral resolution of 4 cm^{-1} with 100% T-line signal-to-noise ratio from 4300 to 4400 cm^{-1}, 64 scans. All measurements were performed with OMNIC software. Samples were ground with potassium bromide at a mass ratio of 100:1 and pressed into tablets. The scanning wavelength was 400–4000 cm^{-1}. The wavelength ratios $R_{1022/995}$ and $R_{1047/1022}$ indicate the short-range of the starch granule's surface, representing the amorphous and crystalline regions of starch, respectively. $R_{1068/1022}$ represents the interaction between protein and starch [43]. The secondary structures of proteins and internal hydrogen bonding were analyzed mainly through the spectral changes in the amide I region (1700 to 1600 cm^{-1}). The main protein structure forms in the amide I region are as follows: β-sheets (1600 to 1640 cm^{-1}), random coils (1640 to 1650 cm^{-1}), α-helixes (1650 to 1660 cm^{-1}), and β-turns (1660 to 1700 cm^{-1}). The changes in these four structures can reflect the changes in the secondary structures of proteins [44].

4.9. Sensory Attributes of Wheat Steamed Bread

Wheat steamed bread was prepared according to the Chinese national standard GB/T 35991-2018 [40]. The sensory attributes and their respective scoring points are specific volume (g/mL, 20 points); width/height ratio (5 points); elasticity (10 points); surface color (10 points); surface structure (10 points); internal structure (20 points); toughness (10 points); viscosity (10 points); and taste (5 points), with a total score of 100. Eight evaluators (five males and three females) aged between 23 and 53 observed, tasted, and evaluated the steamed bread. Each evaluator had a steamed bread sample for each polydextrose addition. The evaluator did not smoke or eat one hour before tasting but could drink water. During the tasting, each evaluator was in a normal physiological state and did not utilize cosmetics or other products with a strong odor. The tasting was carried out one hour before a meal or two hours after a meal.

4.10. Scanning Electron Microscopy (SEM)

The wheat bread samples were cooled to room temperature and kept in a −20 °C refrigerator. Before SEM observation, the samples were lyophilized in a freeze dryer. The dried samples were fixed on a sample stage and then sputtered with gold in a vacuum ion particle sprayer (JEC-3000FC, Japan Electronics Co., Ltd., Tokyo, Japan). The sputtering conditions were as follows: working accelerating voltage of 5.0 kV, sputtering current of 30 mA, sputtering time of 130 s, working distance of 10 mm, and sputtering working pressure of 2.0 Pa. The samples were then set on the stage of a scanning electron microscope (JSM-IT 700HR, Japan Electronics Co., Ltd.) and photographed at an accelerating voltage of 25 kV, with 100 to 1200 times magnification. The pressure in the observation chamber was 7.50×10^{-8} Pa, with an emission current of 88 μA and a distance of 10 mm between the sample and the lens.

4.11. Data Analysis

The data were statistically analyzed using SPSS (Version 17.0, SPSS Incorporated, [45]). For the untreated (CK) and added polydextrose samples, the one-way ANOVA approach and Duncan's new multiple-range test were used to compare multiple means. Statistical significance was set at the $p < 0.05$ level.

Author Contributions: C.L., Data collection, methodology, and writing; B.D., data collection, methodology, and investigation; X.L. (Xiaohong Luo), data collection, and methodology; H.S., supervision and review; X.L. (Xingjun Li), writing—review and editing, methodology, investigation, supervision, results interpretation, and data curation. All authors have read and agreed to the published version of the manuscript.

Funding: This research was supported by the research fund for institute–enterprise cooperation (H24051).

Institutional Review Board Statement: Not applicable.

Informed Consent Statement: The experts who participated in the sensory evaluation of steamed bread consented to the publication of the results.

Data Availability Statement: The original contributions presented in this study are included in this article; further inquiries can be directed to the corresponding author.

Acknowledgments: We thank Guoqing Ning from Runloy Biotech (Shanghai) Co., Ltd., Shanghai, China, for supplying the polydextrose samples.

Conflicts of Interest: The authors declare no conflicts of interest.

References

1. The Society of Chinese Nutrition. *Dietary Nutrient Reference Intakes for Chinese Residents*, 2013rd ed.; Science Press: Beijing, China, 2014; p. 464.
2. Jiang, W.; Zhang, L.C.; Li, W.H.; Chang, C.; Ji, Y.; Li, Y.H. The trend of changes in the level of health literacy for the prevention and treatment of chronic diseases among Chinese residents from 2012 to 2017 and its influencing factors: Age-period-cohort analysis. *China Public Health* **2021**, *37*, 915–920.
3. Saeedi, P.; Petersohn, I.; Salpea, P.; Malanda, B.; Karuranga, S.; Unwin, N.; Colagiuris, S.; Guariguata, L.; Motala, A.A.; Ogurtsova, K.; et al. Global and regional diabetes prevalence estimates for 2019 and projections for 2030 and 2045: Results from the International Diabetes Federation Diabetes Atlas, 9th edition. *Diabetes Res. Clin. Pract.* **2019**, *157*, 107843. [CrossRef] [PubMed]
4. Interpretation of Healthy China Action-Implementation Plan of diabetes Prevention and Control Action (2024–2030). *China Rural Health* **2024**, *16*, 9.

5. Huang, F.F.; Wang, H.J.; Zhang, B.; Wang, Z.H. Update of the China dietary guideline index: CDGI-2022. *Acta Nutr. Sin.* **2024**, *46*, 313–319. [CrossRef]
6. Ansari, F.; Pimentel, T.C.; Pourjafar, H.; Ibrahim, S.A.; Jafari, S.M. The influence of prebiotics on wheat flour, dough, and bread properties; Resistant starch, polydextrose, and inulin. *Foods* **2022**, *11*, 3366. [CrossRef]
7. Canale, M.; Sanfilippo, R.; Strano, M.C.; Bavaro, A.R.; Amenta, M.; Bizzini, M.; Allegra, M.; Blangiforti, S.; Spina, A. Technological properties of inulin-enriched doughs and breads, Influence on short-term storage and glycemic Response. *Foods* **2024**, *13*, 2711. [CrossRef]
8. Topping, D. Cereal complex carbohydrates and their contribution to human health. *J. Cereal Sci.* **2007**, *46*, 220–229. [CrossRef]
9. Yang, Y.Y.; Ma, S.; Wang, X.X.; Zheng, X.L. Modification and application of dietary fiber in foods. *J. Chem.* **2017**, *2017*, 9340427. [CrossRef]
10. Jing, Y.; Wang, W.X.; Yang, L.Z.; Liu, Y.Q.; Shi, M.M. Effect of resistant starch and polydextrose on the quality of steamed bread. *Sci. Technol. Food Ind.* **2020**, *41*, 76–81. [CrossRef]
11. Liu, T.; Duan, H.; Mao, X.; Yu, X. Influence of flaxseed flour as a partial replacement for wheat flour on the characteristics of Chinese steamed bread. *RSC Adv.* **2020**, *10*, 28114–28120. [CrossRef]
12. Kim, J.E.; Baik, B.K.; Park, C.S.; Son, J.H.; Choi, C.H.; Mo, Y.; Park, T.I.; Kang, C.S.; Cho, S.W. Relationship between physicochemical characteristics of Korean wheat flour and quality attributes of steamed bread. *J. Integr. Agric.* **2019**, *18*, 2652–2663. [CrossRef]
13. Dong, X.; Hu, X.; Sun, L.; Zhang, H.; Wu, L.; Wang, B. Volatile compounds of wheat flour and steamed bread as affected by wheat storage time. *SM Anal. Bioanal. Tech.* **2018**, *3*, 1–7. [CrossRef]
14. Staub, H.W.; Mardones, B.; Shah, N. Modern dietary fiber product development and nutrient bioavailability. In *Dietary Fiber*; Birch, G.G., Parker, K.J., Eds.; Applied Science Publishers: New York, NY, USA, 1983; pp. 37–100.
15. Kilibwa, M. Polydextrose As a Fat Absorption Inhibitor in Fried Foods. U.S. Patent No. 6,001,399, 14 December 1999.
16. Guan, Q.Q.; Shao, X.B.; Yang, T.T.; Dou, G.M.; Xu, T.C.; Gan, Z.B. A study on the preparation process of polydextrose. *Agric. Technol.* **2018**, *38*, 247–248.
17. Lauridsen, K. Use of Polydextrose in Edible Products, Edible Products Containing Polydextrose and Processes for Including Polydextrose in Edible Products. U.S. 2004/0213882A1, 28 October 2004.
18. Schirmer, M.; Jekle, M.; Arendt, E.; Becker, T. Physicochemical interactions of polydextrose for sucrose replacement in pound cake. *Food Res. Int.* **2012**, *48*, 291–298. [CrossRef]
19. Guanyan Research Report. The Forecast of the Market Competition Pattern and Development Trend of China Polyglucose Industry on the Analysis Report of 2021 Year. 2021. Available online: https://market.chinabaogao.com/shipin/101455602R021.html (accessed on 14 October 2021).
20. Zhong, J.; Luo, B.Y.; Xiang, M.J.; Liu, H.W.; Zhai, Z.K.; Wang, T.S.; Craig, S.A. Studies on the effects of polydextrose intake on physiologyic functions in Chinese people. *Am. J. Clin. Nutr.* **2000**, *72*, 1503–1509.
21. do Carmo, M.M.R.; Walker, J.C.L.; Novello, D.; Caselato, V.M.; Sgarbient, V.C.; Ouwehand, A.C.; Andreollo, N.A.; Hiane, P.A.; dos Santos, E.F. Polydextrose: Physiological function and effects on health. *Nutrients* **2016**, *8*, 553. [CrossRef]
22. Li, Q.E.; Schmidt, S.J. Use of ramping and equilibrium water-vapor sorption methods to determine the critical relative humidity at which the glassy to rubbery transition occurs in polydextrose. *J. Food Sci.* **2011**, *76*, E149–E157. [CrossRef]
23. Xie, X.H.; Hao, M.Y.; Fan, Y.C.; Zhang, B.; Wang, N.; Guo, F.J. Effect of polydextrose on hydration and structure of gluten during frozen storage. *Food Ferment. Ind.* **2020**, *46*, 78–83. [CrossRef]
24. Guimarães, D.J.S.; Zambelli, R.A.; Afonso, M.R.A.; Pontes, D.F. Cryoprotective potential of vegetable powders and polydextrose in frozen bread doughs. *Int. J. Refrig.* **2025**, *172*, 108–119. [CrossRef]
25. Li, J.T.; Wang, M. Effect of composite powder of chickpea and pea on dough characteristics and the quality of steamed bread. *Food Res. Dev.* **2020**, *41*, 68–72. [CrossRef]
26. Gao, Y.; Liu, T.T.; Su, C.H.; Li, Q.; Yu, X.Z. Fortification of Chinese steamed bread with flaxseed flour and evaluation of its physicochemical and sensory properties. *Food Chem. X* **2022**, *13*, 100267. [CrossRef] [PubMed]
27. Liu, C.; Wang, M.Y.; Song, H.D.; Cao, Z.Y.; Jie, Y.; Li, X.J.; Ning, G.Q. A study on the mechanism of polydextrose moisture absorption analyzed by Hailwood-Horrobin equilibrium moisture model. *Sci. Technol. Cereals Oils Foods* **2024**, *32*, 140–147. [CrossRef]
28. *GB1986.385-2025*; National Food Safety Standard, Food Additive—Polydextrose. China National Standard: Beijing, China, 2025.
29. Rosenthal, A.J. Application of aged egg in enabling increased substitution of sucrose by Litesse (Polydextrose) in high-ratio cakes. *J. Sci. Food Agric.* **1995**, *68*, 127–131. [CrossRef]
30. Jiang, Y. Preliminary evaluation of the application function of polydextrose in bread making. *Tianjin Sci. Technol.* **2009**, *3*, 72–73. [CrossRef]
31. Ficco, D.B.M.; Canale, M.; Giannone, V.; Strano, M.C.; Allegra, M.; Zingale, S.; Spina, A. Durum wheat bread with a potentially high health value through the addition of durum wheat thin bran or barley flour. *Plants* **2023**, *12*, 397. [CrossRef]

32. He, Y.; Guo, J.; Ren, G.; Cui, G.; Han, S.; Liu, J. Effects of konjac glucomannan on the water distribution of frozen dough and corresponding steamed bread quality. *Food Chem.* **2020**, *330*, 127243. [CrossRef]
33. Brasil, J.A.; Silveira, K.C.D.; Salgado, S.M.; Livera, A.V.S.; Faro, Z.P.D.; Guerra, N.B. Effect of the addition of inulin on the nutritional, physical and sensory parameters of bread. *Braz. J. Pharm. Sci.* **2011**, *47*, 185–191. [CrossRef]
34. Wang, M.Y.; Liu, C.; Luo, X.H.; Wu, J.Z.; Li, X.J. Effect of polydextrose on the cooking and gelatinization properties and microstructure of Chinese early indica rice. *Gels* **2025**, *11*, 171. [CrossRef]
35. Pateras, I.M.C.; Howells, K.F.; Rosenthal, A.J. Hot-stage microscopy of cake batter bubbles during simulated baking: Sucrose replacement by polydextrose. *J. Food Sci.* **1994**, *59*, 168–170. [CrossRef]
36. Hicsasmaz, Z.; Yazgan, Y.; Bozoglu, F.; Katnas, Z. Effect of polydextrose-substitution on the cell structure of the high-ratio cake system. *Lebensm. Wiss. Technol.* **2003**, *36*, 441–450. [CrossRef]
37. Kocer, D.; Hicsasmaz, Z.; Bayindirli, A.; Katnas, S. Bubble and pore formation of the high-ratio cake formulation with polydextrose as a sugar- and fat-replacer. *J. Food Eng.* **2007**, *78*, 953–964. [CrossRef]
38. Ding, C.H.; Su, Y.X.; Jiang, Y.M.; Yong, J.Z. Differences in texture, nutritional value and microstructure of steamed buns with different amounts of gelatin. *Mod. Food Sci. Technol.* **2024**, *40*, 274–282. [CrossRef]
39. Seabourn, B.W.; Chung, O.K.; Seib, P.A.; Mathewson, P.R. Determination of secondary structural changes in gluten proteins during mixing using Fourier transform horizontal attenuated total reflectance spectroscopy. *J. Agric. Food Chem.* **2008**, *56*, 4236–4243. [CrossRef] [PubMed]
40. *GB/T 35991–2018*; Inspection of Grain and Oils—Quality Evaluation of Wheat Steamed Bread. China National Standard: Beijing, China, 2018.
41. *GB/T 21118–2007*; Chinese Steamed Bread Made of Wheat Flour. China National Standard: Beijing, China, 2007.
42. *GB/T 24852–2010*; Determination of Pasting Properties of Rice–Rapid Visco Analyzer Method. China National Standard: Beijing, China, 2010.
43. Ha, M.; Jeong, H.Y.; Lim, S.T.; Chung, H.J. The cooking method features controlling eating quality of cooked rice: An explanation from the view of starch structure in leachate and morphological characteristics. *Food Res. Int.* **2022**, *162*, 111980. [CrossRef] [PubMed]
44. Hang, Y.Y.; Zhang, T.T.; Qi, D.; Hu, Y.Q. Effects and mechanism of polydextrose on the freezing-thawing stability of tilapia surimi. *China Food Addit.* **2024**, *10*, 131–138. [CrossRef]
45. SPSS Inc. *SPSS for Windows*, Release 17.0.1; SPSS Inc.: Chicago, IL, USA, 2006.

gels

MDPI

Article

Comparison of Ultrasound- and Microwave-Assisted Extraction Techniques on Chemical, Technological, Rheological, and Microstructural Properties of Starch from Mango Kernel

Luis Mieles-Gómez, Somaris E. Quintana and Luis A. García-Zapateiro *

Research Group on Complex Fluid Engineering and Food Rheology, Universidad de Cartagena, Cartagena 130015, Colombia; lmielesg@unicartagena.edu.co (L.M.-G.); squintanam@unicartagena.edu.co (S.E.Q.)
* Correspondence: lgarciaz@unicartagena.edu.co

Abstract: The effect of emergent technologies for the starch extraction was studied. The evaluation of conventional extraction (MKS-WMP), ultrasound-assisted extraction (MKS-UAE), and microwave-assisted extraction (MKS-MAE) on chemical, technological, gelling, pasting, and microstructural properties of starch from mango kernel was carried out. The extraction yield was found in the values of 42.05, 50.40, and 47.43% for MKS-WMP, MKS-UAE, and MKS-MAE treatments, respectively. The amylose contents for MKS-UAE and MKS-MAE starches were significantly higher ($p < 0.05$) than MKS-WMP, with an increase of about 13–18%. The total phenolic content ranged from 84.89 to 90.85 mg GAE/g starch without significant differences ($p > 0.05$). The technological properties evidence a good water-holding capacity (80.48–90.05 g/100 g of starch) and oil-holding capacity (70.58–83.23 g/100 g of starch). The gelatinization temperature, measured by rheological analysis, ranged between 77 and 82 °C. Microstructural properties showed that ultrasound- and microwave-assisted treatments improved the shape and surface of starch granules, and that they are promising alternatives for starch extraction, providing some characteristics that could increase the applications in the food industry.

Keywords: starch; ultrasound-assisted extraction; microwave-assisted extraction; techno-functional properties; rheological properties

1. Introduction

Mango (*Mangifera indica*) is among the most widely traded tropical fruits globally, ranking second in commercial exchange and fifth in production volume. Recognized as the 'king of fruits', it is consumed at different ripening stages due to its distinctive and appealing flavor [1]. In addition, mango consumption is associated with numerous health benefits due to its composition, including phytochemicals and a rich nutritional profile [2]. Over 50% of the mangoes harvested are processed into various products, such as juice, nectar, puree, jam, and pickles, among others [3]. This generates large quantities of by-products, primarily composed of peels and seeds, which pose environmental challenges for both industry and local communities [4]. The edible portion of the mango fruit varies depending on the variety and accounts for 33–85% of the fresh weight; the seed represents 20–60% of the total fruit mass, of which the kernel comprises 45–75%. The main component of the mango seed kernel is starch, ranging from 38 to 70% [5], which makes it an interesting polysaccharide that can be used from the seed.

Among plant-derived polysaccharides, starch is ranked as the second most abundant after cellulose and plays a key role as an energy source in both food systems and biopolymer applications [6]. It has varied applications in foods (additives such as stabilizers, thickeners, gelling agents, edible films, and encapsulating compounds) and in non-food sectors, including the paper, textile, and cosmetic industries. It is primarily extracted from conventional sources, including cereals and tubers, thus driving demand along the supply chain of these food groups while also impacting costs and environmental sustainability [7]. However, in recent times, a wide variety of new sources of starches have been found that represent an alternative to native starches associated with technological properties similar to or better than those of commercial starches [5]. Various studies have indicated that starch derived from mango seeds exhibits significant potential to compete with commercially relevant starches, such as those from corn, potatoes, and beans, in industrial applications [8,9].

However, to understand and propose an adequate technological application of mango seed starches, it is necessary to explore different extraction techniques to obtain starches with a higher yield rate, greater purity, and desirable functional properties. In this sense, ultrasonic treatment emerges as an effective alternative starch extraction technique. Ultrasonic waves generate cavitation phenomena in liquid media, creating microbubbles that collapse violently and release localized energy. This process disrupts cell walls and facilitates the release of intracellular compounds, such as starch. In starch extraction, ultrasound primarily affects the amorphous regions of the granules, leading to surface erosion, increased porosity, and improved solvent penetration [10,11]. These effects can enhance extraction yield and modify functional properties such as solubility, swelling power, and gelatinization behavior. In addition, it allows for lower consumption of chemical products and shorter extraction times. It also offers great potential for industrial application due to its flexibility to integrate with other processes, low cost, high efficiency, and environmentally friendly profile [10,11]. Likewise, microwaves are non-ionizing electromagnetic waves that generate molecular micromovements and friction in food through high-frequency electric fields, operating based on the dielectric loss principle. Microwave processing can break glycosidic bonds, promote the structural reorganization of starch molecules, and alter the morphology and relative crystallinity of starch granules, leading to modifications in paste viscosity, swelling ability, emulsifying stability, and emulsification properties [12]. Therefore, microwave treatment has gained immense commercial and industrial importance over the years and is being explored for its compatibility with different sources.

Different researchers have used ultrasound and microwave-assisted technology to isolate starch from different botanical sources, such as some legumes and roots [13–15]. Despite this potential, little research has focused on comparing their effect on the technological, rheological, and microstructural properties of mango kernel starch. Consequently, the main objective of this research was to evaluate the effect of ultrasound- and microwave-assisted extraction techniques on mango kernel starch, analyzing their chemical, techno-functional, gelling, pasting, and microstructural properties.

2. Results and Discussion

Mango seed starch (MKS) was obtained using conventional wet-milling process (MKS-WMP), ultrasonic-assisted extraction (MKS-UAE), and microwave-assisted extraction (MKS-MAE), obtaining three starch samples with similar characteristics, that is, a powder with a soft texture and light brown color.

The extraction yield was found in the values of 42.05%, 50.40%, and 47.43% for MKS-WMP, MKS-UAE, and MKS-MAE treatments, respectively (Table 1). Then, the yield extraction of starch by wet milling (conventional method) was close to that reported for mango seed starch of Indian [16] and Brazilian [17] origin, obtained via the wet-milling process, presenting values of 39.35 and 44.95%, respectively. Furthermore, regarding the extraction method, the yield of starches depends on other factors, such as the origin and the condition of the fruit, the requirements of the soil, etc. [18]. However, ultrasound and microwave-assisted extraction presented a significant effect ($p < 0.05$) on starch yield compared to the conventional method. The highest performance value was obtained in the UAE treatment, but without significant differences ($p > 0.05$) with the MAE treatment. These results could be related to the dominant effects of ultrasound cavitation and microwave irradiation. The cavitation effect contributes to increasing the mass transfer of intracellular substances to the solvent, because the particles decompose during sonication, which exposes the structure of the fibrous material of the mango seed to release more starch granules to the outside [19], improving the extraction performance. Likewise, during the MAE process, the solutes are in contact with the solvent, so microwave irradiation can improve the solubility of the compounds and their diffusion from the cells to the solvent, because the plant cells expand and are broken due to the absorption of microwaves, and the active substances of the cells, such as starch, are released and dissolved in the solvent [20]. MAE exhibits good efficiency and shorter extraction time than other extraction techniques. Similar results have been found in obtaining polysaccharides from *Panax notoginseng* by ultrasound and microwave-assisted extraction [21], and microwave/ultrasound-assisted extraction of phytocompounds from the pulp of *Syzygium cumini* [22].

Table 1. Yields (g of dry starch/g of mango kernel flour × 100), amylose content, total phenolic content (TPC), antioxidant activity (AA) of mango kernel starch by wet-milling process (MKS-WMP), ultrasound-assisted extraction (MKS-UAE), and microwave-assisted extraction (MKS-MAE).

Sample Code	Yield %	Amylose Content g/100 g	TPC mg GAE/g	AA μMol Trolox/g
MKS-WMP	42.05 ± 2.58 [b]	28.46 ± 0.93 [b]	84.89 ± 1.55 [a]	18.15 ± 1.10 [b]
MKS-UAE	50.40 ± 2.26 [a]	33.84 ± 1.47 [a]	90.85 ± 3.08 [a]	16.12 ± 1.50 [a]
MKS-MAE	48.43 ± 1.41 [a]	32.21 ± 1.58 [a]	89.60 ± 4.11 [a]	15.24 ± 0.56 [a]

Data are the mean ± standard deviation. Different letters in the same columns express statistically significant differences ($p < 0.05$).

2.1. Amylose Content

Amylose plays a key role in defining the functional behavior of starch, as it influences gelatinization temperature, retrogradation tendency, gel firmness, and overall textural properties in food systems [23]. The amylose content of starch present values between 28.46 and 33.84 g/100 g of starch, as detailed in Table 1. Then, the amylose contents for MKS-UAE and MKS-MAE starches were significantly higher ($p < 0.05$) than MKS-WMP, with an increase of about 13–18%. These results can be attributed to the partial depolymerization of amylose and amylopectin during ultrasonic extraction, which leads to an increased number of linear chains and, consequently, a higher amylose content in starch granules [24]. Similarly, microwave irradiation causes the polar water molecules to vibrate and breakage of the α-1,4 and α-1,6 glycosidic bonds to occur; therefore, the amylose trapped within the granular structure by the amylopectin groups escapes from the semicrystalline structure [25]. The results obtained are consistent with the report that ultrasonic pretreatment increased the amylose content in corn starch at high concentrations [23] and the report that, compared to conventionally extracted starch and without any modification, starch thermally treated with microwave had a higher amylose content [26].

2.2. FTIR Analysis

FTIR spectroscopy has proven to be an effective technique for identifying functional groups in biopolymers, offering valuable insights into structural features such as double-helical arrangement, helicity, and chain conformation within a limited range [27]. The FTIR spectra in the range from 4000 to 550 cm^{-1} of the conventionally extracted starches (WMP), assisted by ultrasound (UAE) and microwaves (MAE), are shown in Figure 1. The FTIR spectrum reveals characteristic peaks at various wavenumbers, reflecting the distinct structural features of the starches. The spectrum pattern did not change, and there was no appearance or loss of new absorption peaks for all samples, demonstrating that no new functional groups or new chemical bonds were formed after the ultrasound- or microwave-assisted extraction technique. Results agree with those reported for pea starch during the process assisted by pre- and post-ultrasound treatment [24] and millet starch treated under microwaves and ultrasound at the same power [28].

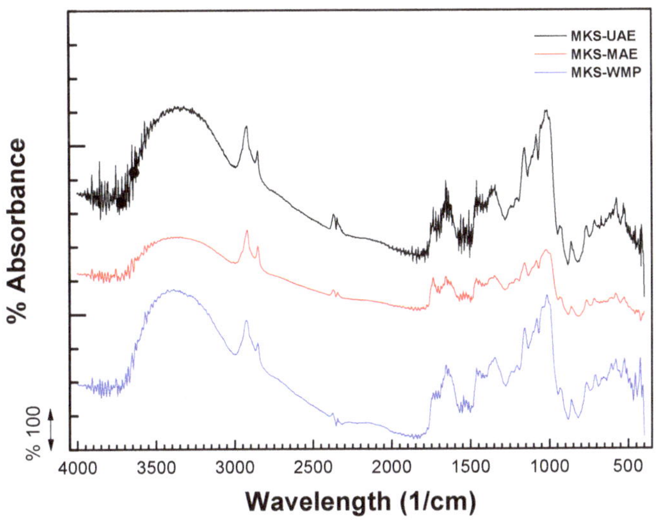

Figure 1. FTIR spectrum of mango kernel starch by wet-milling process (MKS-WMP), ultrasound-assisted extraction (MKS-UAE), and microwave-assisted extraction (MKS-MAE).

The region at 1200–800 cm^{-1} is the band sensitive to the configuration of starch. For all samples, a strong peak was observed around 1022 cm^{-1}; the absorption peak of this part mainly reflects the vibration of C-O, C-C and C-OH bonds and is related to the vibrational absorption of the crystalline regions and starch amorphous. Furthermore, a decrease in the intensity peak was evident in the MKS-MAE sample; this could be attributed to the modification of the starch structure and the development of new intramolecular hydrogen bonds when water evaporates through microwave treatment [29]. Similarly, the peak presented in the wavelength range of 1170–1150 cm^{-1} is the peak of the carbonyl ester, which initially indicates that mango seed starches contain some compounds that are not found in the starch obtained from conventional sources. The peak at 1650 cm^{-1} is related to the bending of water in the amorphous region of starch granules [30]. A broad peak was recorded around 3600–3000 cm^{-1} which represents the stretching vibration of the –OH group of water and starch. The small peak at 2900 cm^{-1} is attributed to C-H stretching of aliphatic groups, reflecting the presence of small amounts of lipids [31].

2.3. Total Phenolic Compounds (TPC) and Antioxidant Capacity

The determined TPC and the antioxidant capacity of starches are shown in Table 1. The TPC content of the starch samples ranged from 84.89 to 90.85 mg GAE/g starch without significant differences ($p > 0.05$); that is, the extraction assisted by ultrasound and microwaves did not influence the total phenolic content, suggesting that polyphenols are strongly linked to starch chains, supporting the effect of cavitation, and ultrasound and microwave irradiation, respectively. Similarly, the results of antioxidant activity indicated that ultrasound and microwave significantly reduced ($p < 0.05$) compared to the conventionally extracted sample; this effect may be associated with localized thermal degradation and the generation of hydroxyl radicals induced by bubble formation during microwave and ultrasound treatments. During cavitation, the presence of these hydroxyl radicals in water can lead to the oxidation of certain polar compounds [32].

2.4. Technological Properties

2.4.1. Water-Holding Capacity (WHC)

The water absorption capacity determines the weight of water that can be retained per unit weight of dry starch. The water-retention capacity (WHC) was expressed in g per 100 g of starch. The WHC ranged between 80.48 and 90.05 g/100 g of starch, as shown in Table 2. The WHC values for the MKS-UAE and MKS-MAE starches were significantly higher ($p < 0.05$) than for MKS-WMP starch. The increase in WHC in the sonicated sample could be explained by the cavitation-driven decomposition of starch molecules into small-sized granules, exhibiting enhanced specific exposed granule areas compared to those of conventionally extracted starch molecules [33]. Similarly, microwave-assisted extraction contributes to the increase in the specific net energy absorbed by starch, favoring the alteration of the morphological (crack formation) and molecular structure of the starch granules, followed by the entry of water into the interior regions. During microwave treatment, the removal of moisture causes severe tension between crystals, leading to transformation of starch nature from crystalline to amorphous and allowing unhindered reaction with water, which is represented as an increase in WHC [29]. These results are in agreement with the findings of Falsafi et al. [34] for sonicated oat samples that demonstrated a significant increase ($p < 0.05$) in WHC when compared to the non-sonicated sample, and the results obtained by Gani et al. [35] for starch from lotus stem harvested from Dal Lake of Jammu and Kashmir, India, with microwave application.

Table 2. Water-holding capacity (WHC), oil-holding capacity (OHC), solubility, and swelling power (SP) of mango kernel starch by wet-milling process (MKS-WMP), ultrasound-assisted extraction (MKS-UAE), and microwave-assisted extraction (MKS-MAE).

Sample Code	WHC g/100 g	OHC g/100 g	Solubility			Swelling Power		
			25 °C	65 °C	90 °C	25 °C	65 °C	90 °C
MKS-WMP	80.48 ± 2.41 [b]	76.43 ± 1.63 [b]	1.10 ± 0.07 [c]	6.20 ± 0.15 [c]	15.00 ± 0.58 [b]	1.97 ± 0.09 [b]	6.17 ± 0.04 [a]	14.88 ± 0.47 [b]
MKS-UAE	87.76 ± 2.01 [a]	83.23 ± 1.83 [a]	1.52 ± 0.04 [a]	6.86 ± 0.08 [a]	16.75 ± 0.30 [a]	2.74 ± 0.32 [a]	6.32 ± 1.02 [a]	12.89 ± 0.85 [a]
MKS-MAE	90.05 ± 2.54 [a]	78.56 ± 1.32 [a]	1.34 ± 0.05 [b]	6.46 ± 0.12 [b]	16.88 ± 0.38 [a]	2.49 ± 0.46 [ab]	6.45 ± 0.74 [a]	15.72 ± 1.10 [b]

Data are the mean ± standard deviation. Different letters in the same columns express statistically significant differences ($p < 0.05$).

2.4.2. Oil-Holding Capacity (OHC)

OHC is directly related to flavor retention and mouthfeel, making it an important parameter to consider. The OHC of MKS-WMP starch was 76.43 g/100 g of starch, which was reduced to 70.58 g/100 g after microwave-assisted extraction. This phenomenon may be attributed to a reduction in the number of hydrophobic sites or an increased tendency of starch to resist interaction with oil during treatment, as reported by Nadir et al. [36] in potato starch with microwave treatment. In contrast, ultrasound-assisted extraction caused an increase in OHC to 83.23 g/100 g; this increase is mainly due to the destruction of the starch structure by the cavitation effect, leading to greater capillary trapping of molecules of oil in the empty fraction and the filling of interstitial spaces between the starch granules. These findings are in agreement with those reported for potato and yam starch in studies where the impact of sonication on the aggregation structure and physicochemical characteristics was evaluated [37,38]. These findings indicate that ultrasound-assisted extraction enhances the emulsifying properties of MKS, broadening its potential applications in various food formulations.

2.4.3. Solubility and Swelling Power (SP)

The solubility and swelling capacity values were evaluated at temperatures of 25, 65, and 90 °C, as shown in Table 2. In all cases, the SP values increased gradually with increasing temperature, which can be attributed to the weakening of hydrogen bonds between intermolecular interactions at high temperatures, further aggravating the swelling of the samples. In the case of ultrasound-assisted extraction, there were only significant differences ($p < 0.05$) at a temperature of 90 °C, which showed a decrease in the SP value when compared to MKS-WMP and MKS-MAE, possibly due to depolymerization of starch molecules caused by ultrasound; this leads to the breakdown of the amylopectin chain. Additionally, at elevated temperatures, starch undergoes gelatinization, and the resulting gelatinized starch, along with the denatured protein matrix, can hinder water penetration into the starch structure [39]. Similar results were found in pea starch extracted with ultrasound assistance and lentil starch modified by ultrasound [24,40]. However, microwave-assisted extraction did not modify the SP values when compared to the conventionally obtained sample. These results are not in accordance with those reported by Palay [41], who pointed out that microwave treatment decreased the swelling capacity and was due to gelatinization by microwave irradiation and the absence of a continuous network of amylose chains.

The solubility percentage of mango seed starch increased markedly with increasing temperature from 25 to 90 °C, which may be due to increased leaching of amylose with increasing temperature ($p < 0.05$). The increase in MKS-MAE starch solubility may also be correlated with the variation induced by microwave irradiation in the starch structure. Throughout this process, the movement of polar molecules, along with unavoidable friction, dipole rotation, and molecular collisions, enhances the functional properties of starch, particularly its solubility. Results that coincide with ours were recently reported for starch isolated from lotus seed flour (*Nelumbo nucifera*) thermally treated with microwaves [42]. The present results corroborate those from previous reports on the influence of microwave treatment on the functional properties of starches and seed flours [42–44].

2.5. Rheological Properties

The gelatinization, defined as the intersection corresponding to the intersection of G' (storage modulus) and G'' (loss modulus) [45] of samples, is shown in Figure 2. It was observed that, at the beginning of heating, within the range from 40 to 75 °C, the values of G' were lower than those of G'' for all samples and were basically stable. After 75 °C, G' and G'' increased sharply. As the temperature increased, the two modules crossed ($G' = G''$), indicating the start temperature of gelatinization of the studied starches.

The gelatinization temperature values measured during the rheological tests were 82.68, 77.52, and 81.70 °C for the MKS-UAE, MKS-MAE, and MKS-WMP samples, respectively. As the temperature increased, starch granules absorbed water, causing them to swell, while amylose was released and interconnected to create a network structure. However, further heating resulted in the melting, fragmentation, and breakdown of the starch granules, along with a weakening of the intermolecular interactions, so the G' values were higher than those of G'', showing a different trend than the previous one. Similar behaviors have been found in different starch suspensions, such as potato, cassava [46], and Peruvian carrot starch peruana [47]. The gelatinization temperature of MKS-MAE starch was lower and presented significant differences from the rest of the samples ($p > 0.05$), This effect may be attributed to microwave irradiation, which significantly disrupted both internal and intermolecular hydrogen bonds within the starch structure. As a result, starch chains became more exposed, leading to a reduction in amylopectin content. Consequently, water molecules could infiltrate the starch granules more efficiently, facilitating gelatinization and lowering the gelatinization temperature [48]. Similarly, MKS-UAE starch maintained comparable gelatinization characteristics to the conventional sample (MKS-WMP), suggesting that ultrasound treatment did not significantly alter the overall crystalline structure or energy required for gelatinization. However, structural modifications observed in SEM images—such as surface fissures—could still enhance hydration and influence functional behavior in specific applications.

From an industrial perspective, the lower gelatinization temperature of MKS-MAE starch is advantageous for applications in thermally sensitive food products, such as instant puddings, sauces, or powdered beverages, for which rapid solubilization at lower temperatures is desired [49].

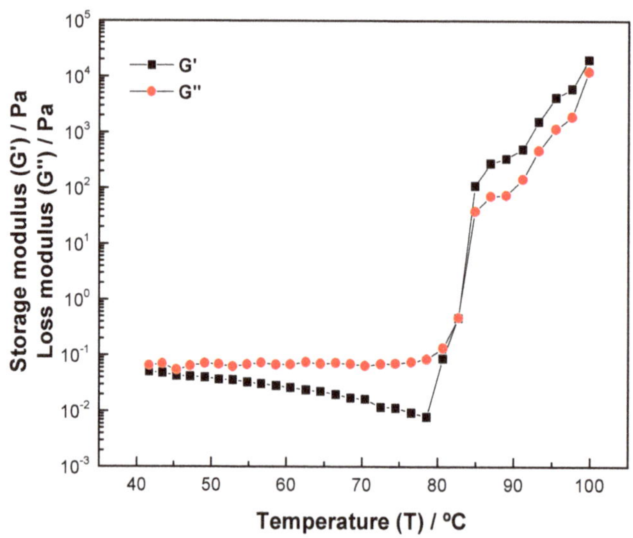

(a)

Figure 2. *Cont.*

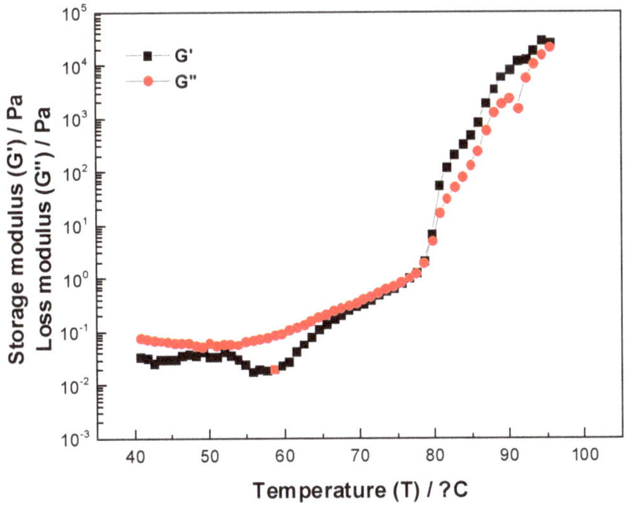

(b)

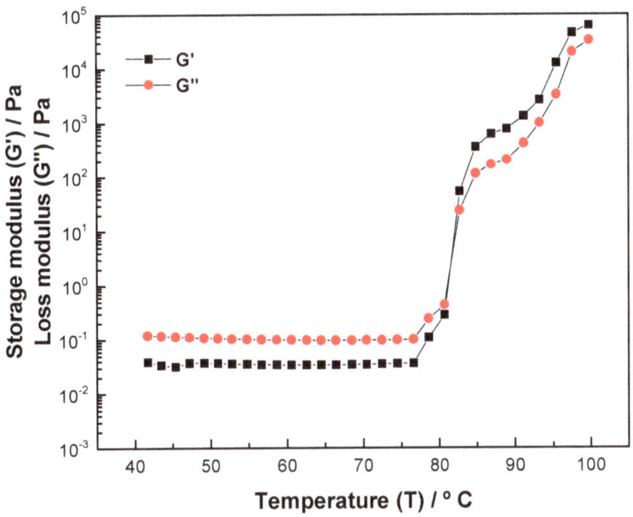

(c)

Figure 2. The storage (G′) and loss (G″) moduli of mango kernel starch by (**a**) wet-milling process (MKS-WMP), (**b**) ultrasound-assisted extraction (MKS-UAE), and (**c**) microwave-assisted extraction (MKS-MAE).

2.6. Pasting Properties

The profiles of the bonding property curves are shown in Figure 3, and the bonding parameters of the mango seed starches extracted with the assistance of ultrasound, with the assistance of microwaves, and conventionally are summarized in Table 3.

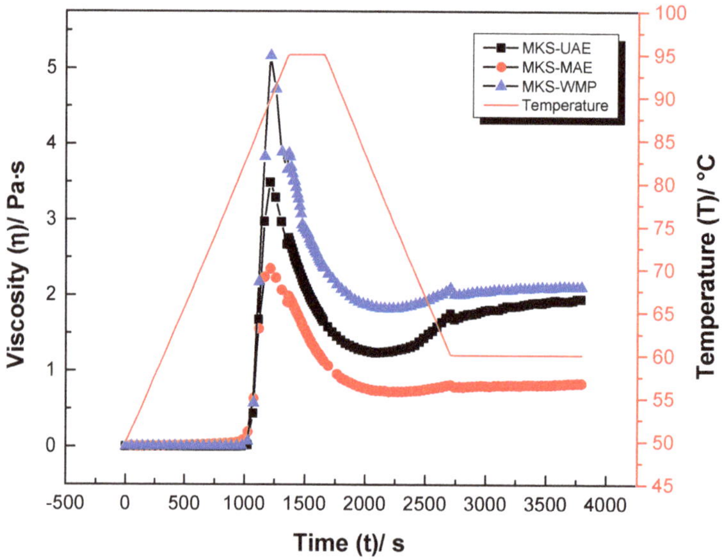

Figure 3. Pasting properties of mango kernel starch by wet-milling process (MKS-WMP), ultrasound-assisted extraction (MKS-UAE), and microwave-assisted extraction (MKS-MAE).

Table 3. Pasting properties of mango kernel starch by wet-milling process (MKS-WMP), ultrasound-assisted extraction (MKS-UAE), and microwave-assisted extraction (MKS-MAE). PT, pasting temperature; PV, peak viscosity; TV, trough viscosity; BD, breakdown viscosity; FV, final viscosity; SB, setback viscosity.

Sample Code	PT °C	PV Pa·s	FV Pa·s	TV Pa·s	BV Pa·s	SB Pa·s
MKS-WMP	83.80 ± 1.67 ^{b}	5.17 ± 0.25 ^{c}	2.11 ± 0.10 ^{c}	1.83 ± 0.09 ^{c}	3.34 ± 0.16 ^{c}	3.06 ± 0.15 ^{b}
MKS-UAE	82.84 ± 1.47 ^{ab}	3.49 ± 0.19 ^{a}	1.94 ± 0.07 ^{a}	1.23 ± 0.08 ^{a}	2.26 ± 0.15 ^{a}	1.54 ± 0.05 ^{a}
MKS-MAE	79.85 ± 1.41 ^{a}	2.35 ± 0.12 ^{b}	0.83 ± 0.05 ^{b}	0.72 ± 0.09 ^{b}	1.63 ± 0.11 ^{b}	1.52 ± 0.10 ^{a}

Data are the mean ± standard deviation. Different letters in the same columns express statistically significant differences ($p < 0.05$).

The PT of the MKS-WMP sample was 83.80 °C, higher than that of samples extracted with ultrasound and microwaves. Additionally, ultrasound and microwave treatments may modify the crystalline structure of starch granules, disrupt macromolecular chains, and decrease the granules' rigidity and structural integrity, ultimately leading to a reduction in maximum viscosity [50]. The lower viscosity could be attributed to the partial breakdown of the integrity of the starch granules, which in turn led to a greater penetration of water for hydration, results that agree with the results obtained for the water-retention capacity of the samples MKS-UAE and MKS-MAE. Furthermore, Studies have indicated that the mechanical energy generated by ultrasound and cavitation can fragment long molecular chains and weaken the interaction forces between starch granules, ultimately resulting in a decrease in the peak viscosity [51].

The breakdown viscosity of the MKS-UAE, MKS-MAE, and MKS-WMP samples was found to be 2.26, 1.63, and 3.34 Pa·s, respectively. Consequently, the MKS-MAE sample exhibited the highest resistance to shear thinning under thermal processing, likely due to the degradation of glycosidic bonds induced by the vibration of polar molecules during microwave treatment. A similar behavior has been reported for chestnut starch subjected to microwave processing [52]. The low setback values reflect the stability of starch against deterioration and aging. In the present study, the setback viscosity values decreased significantly ($p < 0.05$) with the application of ultrasound and microwaves, and this effect may be attributed to the disruption of amylose rearrangement caused by ultrasonic treatment; microwave processing also contributed positively to this modification [53].

2.7. Scanning Electron Microscopy (SEM)

SEM micrographs of sample starches at 1000× and 3200× magnification are presented in Figure 4a–f. The extraction method influenced the surface morphology and granule integrity. In general, both UAE and MAE treatments led to a reduction in granule size compared to the conventional method. Also, MKS-WMP exhibits a distinct angular shape with an irregular surface (Figure 4a,d), and some granules show smooth surfaces, indicating partial preservation of native morphology, whereas MKS-UAE (Figure 4b,e) has an oval shape with agglomerate, a smooth surface, and small fissure depressions. These features are associated with the selective disruption of the amorphous regions of starch granules by ultrasound cavitation [54], which induces erosion and physical disintegration without completely collapsing the granule structure [55]. MKS-MAE (Figure 4c,f) presents irregular oval granules with more pronounced cracks, fissures, and surface scratches. The intense molecular movement induced by microwave energy may cause thermal stress, localized overheating, and fragmentation, resulting in granule deformation and structural collapse [26,56]. The higher degree of surface disruption observed in MAE samples could explain their higher solubility and water-holding capacity, as more porous structures allow for greater water penetration and starch chain mobility.

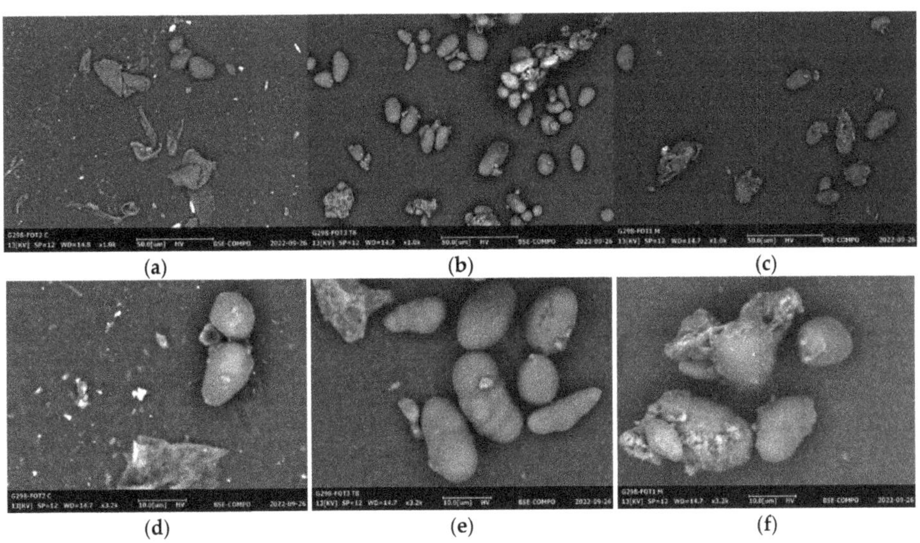

(a) (b) (c)

(d) (e) (f)

Figure 4. SEM micrographs of mango kernel starch by (**a**) wet-milling process (MKS-WMP), (**b**) ultrasound-assisted extraction (MKS-UAE), and (**c**) microwave-assisted extraction (MKS-MAE) at 1000X; and (**d**) MKS-WMP, (**e**) MKS-UAE, and (**f**) MKS-MAE at 3200X.

These microstructural differences are consistent with the technological properties discussed earlier. The disrupted surfaces observed in UAE and MAE starches may enhance their interaction with water and oil, influencing swelling, emulsification, and viscosity behavior—key attributes in food-formulation applications.

3. Conclusions

Starch extraction by UAE and MAE exhibited a significantly high extraction yield compared to the WMP method; in addition, the UAE and MAE technologies increased amylose content. The FTIR spectrum did not change, suggesting that ultrasound and microwave treatments exhibited solely physical effects on the starch samples; in addition, the UAE and MAE did not influence the total phenolic content. The technofunctional properties of WFC, OHC, solubility, and SP were improved through ultrasound- and microwave-assisted extraction, allowing for the expansion of these starches in different industrial applications. Extraction by MAE made it possible to accelerate the gelatinization process and reduce the gelatinization temperature. Ultrasound and microwave treatment decrease the size of mango kernel starch, improving the surface and shape. The results of this research indicated that UAE and MAE provide some novel characteristics to mango seed starch that contribute to increasing the applications of these starches in the food industry, such as instant powder type, meat industry, or stabilizers of microstructured products.

4. Materials and Methods

4.1. Materials

Anhydrous sodium carbonate (99.5% purity) and sodium bisulfite (99.5% purity) were purchased from Sigma-Aldrich (St. Louis, MO, USA). Ethanol (99.5% purity) was purchased from Panreac (Barcelona, Spain). Mango (*Mangifera indica*) var. Coraz6n fruits were harvested to commercial maturity.

4.2. Obtention of Mango Kernel Flour

In order to separate de kernel, the mango seed was washed and dried at 40 °C for 12 h in a hot air -tray dryer and milled to obtain a flour with a particle size less than 250 μm.

4.3. Wet-Milling Process of Mango Kernel Starch

The obtention of mango kernel starch by wet-milling process (MKS-WMP) was carried out following the methodology proposed by Kaur et al. [8], with some modifications. A suspension of 10 g of flour was made in 90 mL of aqueous sodium bisulfite solution (1%) and was stirred for 4 h at room temperature. Subsequently, the homogenization was carried out on an Ultra-Turrax (IKA T25, Staufen, Germany) at 650 rpm × 5 min in order to ensure that the raw material was completely solubilized. The mixture was filtered, and the cake was washed with distilled water. Then, the filtrate was centrifuged at 5000 rpm for 10 min, and the precipitate was dried at 40 °C until constant weight. The extraction yield of starch was calculated based on the initial dry weight amount of mango flour (mango kernel flour) and obtained starch (dry starch), as expressed in Equation (1).

$$\text{Yield } (\%) = \frac{\text{Dry starch } (g)}{\text{Mango kernel flour}(g)} \times 100 \quad (1)$$

4.4. Ultrasound-Assisted Extraction (UAE) of Mango Kernel Starch

The obtention of mango kernel starch by ultrasound-assisted extraction (MKS-UAE) was carried out following the procedure described by Mieles et al. [57]. The procedures for wet-milling process of mango kernel starch applying an ultrasound to the suspensions in solubilization step using an Ultrasonic Processor (SX Sonic, FS-1200N, Shanghai, China) with a 20 mm diameter probe that was submerged at a depth of 50 mm from the surface of the suspension, operating at 360 W for 10 min at 40 °C and constant frequency of 20 kHz, with pulse durations of 2 s on and 2 s off.

4.5. Microwave-Assisted Extraction (MAE) of Mango Kernel Starch

The obtention of mango kernel starch by microwave-assisted extraction (MKS-MAE) was carried out using the methodology described by Ramírez-Brewer et al. [58], with some modifications. The procedure for the wet-milling process of mango kernel starch involves applying microwaves on the suspensions in a solubilization step for a period of 90 s, using a microwave oven at a low–medium power of 360 W.

4.6. Determination of the Amylose and Amylopectin Content

The amylose content was determined through the iodine method, following the procedure described by Morrison and Laignelet [59].

4.7. Fourier Transform Infrared Spectroscopy (FTIR) Analysis

Fourier-transform infrared spectroscopy (FTIR) in a medium spectrum range was performed in order to identify the functional groups of starch, employing an FTIR spectrometer (Shimadzu model IR-Affinity, Kyoto, Japan). The samples were combined with KBr at a 1:100 (w/w) ratio and dried, and then their IR spectra were recorded within the 400–4000 cm^{-1} wavelength range, with a resolution of 4 cm^{-1}, at room temperature. The intensity of the spectra was measured by recording the height of the absorbance bands relative to the baseline.

4.8. Determination of Total Phenolic Compounds (TPC) and Antioxidant Capacity

The total phenolic content of samples was determined using the Folin–Ciocâlteu method [60], and the results were expressed as mg gallic acid equivalents (GAE)/g starch. The antioxidant capacity was measured following the method described by Re et al. [61], using the ABTS free radical scavenging test. The results were expressed as μmol of Trolox (TEAC)/g of starch.

4.9. Technological Properties Analysis

The water-holding capacity (WHC) and oil-holding capacity (OHC) of the samples were evaluated using the methodology outlined by Yamazaki [62], modified by Medcalf [63].

The solubility (weight of dried supernatant/weight of sample × 100) and swelling power (weight of wet sediment/weight of sample (100%-% solubility) × 100) of the starch samples were assessed following the methodology described by Jiang et al. [64].

4.10. Rheological Analysis

Starch was mixed with distilled water in order to obtain suspensions at 15% w/w. The suspensions were subjected to rheological analysis, using the procedure established by López-Barraza et al. [65]: employ a controlled stress rheometer (Modular Advanced Rheometer System Haake Mars 60, Thermo-Scientific, Dreieich, Germany) equipped with a rough plate geometry (35 mm diameter and GAP diameter) to prevent wall slip effects.

Stress sweep: In order to identify the linear viscoelastic region (LVR), a stress amplitude sweep test of samples was performed within a range from 0.001 to 1000 Pa, at angular frequencies of 0.01, 0.1, and 1 Hz at 25 °C.

Temperature sweep: The temperature sweep was conducted within the LVR from 40 to 90 °C, using a heating rate of 5 °C/min at a constant frequency of 1 Hz.

Pasting properties: The temperature profile was determined following the methodology described by Mieles et al. [58]: briefly, heating ramps start from 50 to 95 °C at 5 °C/min, hold at 95 °C for 5 min, cool down from 95 to 60 °C at 5 °C/min, and finally kept at 60 °C for 5 min.

4.11. Microstructural Analysis

Microstructural analysis was performed by employing a scanning electron microscope (JSM-6390LV, JEOL, Ltd., Tokyo, Japan) operating at 5 keV. The samples were affixed to aluminum stubs using double-sided carbon tape and coated with a thin gold layer through ion sputtering under vacuum conditions.

4.12. Statistical Analysis

Data were analyzed with a unidirectional ANOVA, using Statgraphics software (version centurión XVI), in order to determine statistically significant differences ($p < 0.05$) between samples. The Least Significant Difference (LSD) test was used for the comparison of means. All tests were performed in triplicate.

Author Contributions: Conceptualization, L.M.-G., S.E.Q. and L.A.G.-Z.; methodology, L.M.-G., S.E.Q. and L.A.G.-Z.; software, L.M.-G. and S.E.Q.; validation, L.M.-G., S.E.Q. and L.A.G.-Z.; formal analysis, L.M.-G. and S.E.Q.; investigation, L.M.-G.; resources, L.M.-G., S.E.Q. and L.A.G.-Z.; data curation, L.M.-G. and S.E.Q.; writing—original draft preparation, L.M.-G. and S.E.Q.; writing—review and editing, L.M.-G., S.E.Q. and L.A.G.-Z.; visualization, L.M.-G., S.E.Q. and L.A.G.-Z.; supervision, L.M.-G., S.E.Q. and L.A.G.-Z.; project administration, L.M.-G., S.E.Q. and L.A.G.-Z.; funding acquisition, L.M.-G., S.E.Q. and L.A.G.-Z. All authors have read and agreed to the published version of the manuscript.

Funding: This work was supported by the Ministry of Science, Technology, and Innovation—MinCiencias (Contract: 368–2019; research project No. 110780864755); and the Strengthening Plan for research groups from University of Cartagena-Resolution 044-2021.

Institutional Review Board Statement: Not Applicable.

Informed Consent Statement: Not Applicable.

Data Availability Statement: The original contributions presented in this study are included in the article. Further inquiries can be directed to the corresponding author.

Acknowledgments: This work is part of a research program 'Programa Nacional de Ciencia, Tecnologia y Innovación en Ciencias Agropecuarias; Project 368-2019 code 110780864755' sponsored by MinCiencias (Colombia). The authors are grateful for their financial support.

Conflicts of Interest: The authors declare no conflicts of interest.

References

1. FAO. Production of Mango, Mangosteens and Guava in India. Available online: http://www.fao.org/faostat/en/#data/QC (accessed on 27 March 2023).
2. Torres-León, C.; Rojas, R.; Contreras-Esquivel, J.C.; Serna-Cock, L.; Belmares-Cerda, R.E.; Aguilar, C.N. Mango Seed: Functional and Nutritional Properties. *Trends Food Sci. Technol.* **2016**, *55*, 109–117. [CrossRef]
3. Nadeem, M.; Imran, M.; Khalique, A. Promising Features of Mango (*Mangifera indica* L.) Kernel Oil: A Review. *J. Food Sci. Technol.* **2016**, *53*, 2185–2195. [CrossRef]
4. Mwaurah, P.W.; Kumar, S.; Kumar, N.; Panghal, A.; Attkan, A.K.; Singh, V.K.; Garg, M.K. Physicochemical Characteristics, Bioactive Compounds and Industrial Applications of Mango Kernel and Its Products: A Review. *Compr. Rev. Food Sci. Food Saf.* **2020**, *19*, 2421–2446. [CrossRef] [PubMed]
5. Punia Bangar, S.; Kumar, M.; Whiteside, W.S. Mango Seed Starch: A Sustainable and Eco-Friendly Alternative to Increasing Industrial Requirements. *Int. J. Biol. Macromol.* **2021**, *183*, 1807–1817. [CrossRef] [PubMed]
6. Gusmão, T.A.S.; de Gusmão, R.P.; Moura, H.V.; Silva, H.A.; Cavalcanti-Mata, M.E.R.M.; Duarte, M.E.M. Production of Prebiotic Gluten-Free Bread with Red Rice Flour and Different Microbial Transglutaminase Concentrations: Modeling, Sensory and Multivariate Data Analysis. *J. Food Sci. Technol.* **2019**, *56*, 2949–2958. [CrossRef]
7. Tagliapietra, B.L.; Felisberto, M.H.F.; Sanches, E.A.; Campelo, P.H.; Clerici, M.T.P.S. Non-Conventional Starch Sources. *Curr. Opin. Food Sci.* **2021**, *39*, 93–102. [CrossRef]

8. Kaur, M.; Singh, N.; Sandhu, K.S.; Guraya, H.S. Physicochemical, Morphological, Thermal and Rheological Properties of Starches Separated from Kernels of Some Indian Mango Cultivars (*Mangifera indica* L.). *Food Chem.* **2004**, *85*, 131–140. [CrossRef]
9. Nawab, A.; Alam, F.; Haq, M.A.; Hasnain, A. Effect of Guar and Xanthan Gums on Functional Properties of Mango (Mangifera Indica) Kernel Starch. *Int. J. Biol. Macromol.* **2016**, *93*, 630–635. [CrossRef]
10. Chang, Y.; Yan, X.; Wang, Q.; Ren, L.; Tong, J.; Zhou, J. High Efficiency and Low Cost Preparation of Size Controlled Starch Nanoparticles through Ultrasonic Treatment and Precipitation. *Food Chem.* **2017**, *227*, 369–375. [CrossRef]
11. Maniglia, B.C.; Castanha, N.; Rojas, M.L.; Augusto, P.E. Emerging Technologies to Enhance Starch Performance. *Curr. Opin. Food Sci.* **2021**, *37*, 26–36. [CrossRef]
12. Zhong, Y.; Liang, W.; Pu, H.; Blennow, A.; Liu, X.; Guo, D. Short-Time Microwave Treatment Affects the Multi-Scale Structure and Digestive Properties of High-Amylose Maize Starch. *Int. J. Biol. Macromol.* **2019**, *137*, 870–877. [CrossRef] [PubMed]
13. Karaman, M.; Tuncel, N.B.; Yılmaz Tuncel, N. The Effect of Ultrasound-Assisted Extraction on Yield and Properties of Some Pulse Starches. *Starch—Stärke* **2017**, *69*, 1600307. [CrossRef]
14. Li, Y.; Wu, Z.; Wan, N.; Wang, X.; Yang, M. Extraction of High-Amylose Starch from Radix Puerariae Using High-Intensity Low-Frequency Ultrasound. *Ultrason. Sonochem.* **2019**, *59*, 104710. [CrossRef] [PubMed]
15. González-Lemus, L.B.; Calderón-Domínguez, G.; Salgado-Cruz, M.d.l.P.; Díaz-Ramírez, M.; Ramírez-Miranda, M.; Chanona-Pérez, J.J.; Güemes-Vera, N.; Farrera-Rebollo, R.R. Ultrasound-Assisted Extraction of Starch from Frozen Jicama (*P. erosus*) Roots: Effect on Yield, Structural Characteristics and Thermal Properties. *CyTA—J. Food* **2018**, *16*, 738–746. [CrossRef]
16. Bharti, I.; Singh, S.; Saxena, D.C. Exploring the Influence of Heat Moisture Treatment on Physicochemical, Pasting, Structural and Morphological Properties of Mango Kernel Starches from Indian Cultivars. *LWT* **2019**, *110*, 197–206. [CrossRef]
17. Ferraz, C.A.; Fontes, R.L.S.; Fontes-Sant'Ana, G.C.; Calado, V.; López, E.O.; Rocha-Leão, M.H.M. Extraction, Modification, and Chemical, Thermal and Morphological Characterization of Starch From the Agro-Industrial Residue of Mango (*Mangifera indica* L) Var. Ubá. *Starch—Stärke* **2019**, *71*, 1800023. [CrossRef]
18. Tesfaye, T.; Johakimu, J.K.; Chavan, R.B.; Sithole, B.; Ramjugernath, D. Valorisation of Mango Seed via Extraction of Starch: Preliminary Techno-Economic Analysis. *Clean Technol. Environ. Policy* **2018**, *20*, 81–94. [CrossRef]
19. Dey, S.; Rathod, V.K. Ultrasound Assisted Extraction of β-Carotene from Spirulina Platensis. *Ultrason. Sonochem.* **2013**, *20*, 271–276. [CrossRef]
20. Elakremi, M.; Sillero, L.; Ayed, L.; ben Mosbah, M.; Labidi, J.; ben Salem, R.; Moussaoui, Y.; Pistacia Vera, L. Leaves as a Renewable Source of Bioactive Compounds via Microwave Assisted Extraction. *Sustain. Chem. Pharm.* **2022**, *29*, 100815. [CrossRef]
21. Shen, S.; Zhou, C.; Zeng, Y.; Zhang, H.; Hossen, M.A.; Dai, J.; Li, S.; Qin, W.; Liu, Y. Structures, Physicochemical and Bioactive Properties of Polysaccharides Extracted from Panax Notoginseng Using Ultrasonic/Microwave-Assisted Extraction. *LWT* **2022**, *154*, 112446. [CrossRef]
22. Sharma, M.; Dash, K.K. Microwave and Ultrasound Assisted Extraction of Phytocompounds from Black Jamun Pulp: Kinetic and Thermodynamics Characteristics. *Innov. Food Sci. Emerg. Technol.* **2022**, *75*, 102913. [CrossRef]
23. Li, C.; Liu, W.; Gu, Z.; Fang, D.; Hong, Y.; Cheng, L.; Li, Z. Ultrasonic Pretreatment Improves the High-Temperature Liquefaction of Corn Starch at High Concentrations. *Starch—Stärke* **2017**, *69*, 1600002. [CrossRef]
24. Han, L.; Cao, S.; Yu, Y.; Xu, X.; Cao, X.; Chen, W. Modification in Physicochemical, Structural and Digestive Properties of Pea Starch during Heat-Moisture Process Assisted by Pre- and Post-Treatment of Ultrasound. *Food Chem.* **2021**, *360*, 129929. [CrossRef] [PubMed]
25. Seung, D. Amylose in Starch: Towards an Understanding of Biosynthesis, Structure and Function. *New Phytol.* **2020**, *228*, 1490–1504. [CrossRef] [PubMed]
26. Wang, M.; Liu, G.; Li, J.; Wang, W.; Hu, A.; Zheng, J. Structural and Physicochemical Properties of Resistant Starch under Combined Treatments of Ultrasound, Microwave, and Enzyme. *Int. J. Biol. Macromol.* **2023**, *232*, 123331. [CrossRef]
27. Sun, H.; Fan, J.; Tian, Z.; Ma, L.; Meng, Y.; Yang, Z.; Zeng, X.; Liu, X.; Kang, L.; Nan, X. Effects of Treatment Methods on the Formation of Resistant Starch in Purple Sweet Potato. *Food Chem.* **2022**, *367*, 130580. [CrossRef]
28. Li, Y.; Hu, A.; Zheng, J.; Wang, X. Comparative Studies on Structure and Physiochemical Changes of Millet Starch under Microwave and Ultrasound at the Same Power. *Int. J. Biol. Macromol.* **2019**, *141*, 76–84. [CrossRef] [PubMed]
29. Kumar, Y.; Singh, L.; Sharanagat, V.S.; Patel, A.; Kumar, K. Effect of Microwave Treatment (Low Power and Varying Time) on Potato Starch: Microstructure, Thermo-Functional, Pasting and Rheological Properties. *Int. J. Biol. Macromol.* **2020**, *155*, 27–35. [CrossRef]
30. Zhou, D.; Ma, Z.; Yin, X.; Hu, X.; Boye, J.I. Structural Characteristics and Physicochemical Properties of Field Pea Starch Modified by Physical, Enzymatic, and Acid Treatments. *Food Hydrocoll.* **2019**, *93*, 386–394. [CrossRef]
31. Morales-Trejo, F.; Trujillo-Ramírez, D.; Aguirre-Mandujano, E.; Lobato-Calleros, C.; Vernon-Carter, E.J.; Alvarez-Ramirez, J. Ultrasound-Assisted Extraction of Lychee (*Litchi chinensis* Sonn.) Seed Starch: Physicochemical and Functional Properties. *Starch—Stärke* **2022**, *74*, 2100092. [CrossRef]

32. Sivasankar, T.; Paunikar, A.W.; Moholkar, V.S. Mechanistic Approach to Enhancement of the Yield of a Sonochemical Reaction. *AIChE J.* **2007**, *53*, 1132–1143. [CrossRef]
33. Raza, H.; Ameer, K.; Ma, H.; Liang, Q.; Ren, X. Structural and Physicochemical Characterization of Modified Starch from Arrowhead Tuber (*Sagittaria sagittifolia* L.) Using Tri-Frequency Power Ultrasound. *Ultrason. Sonochem.* **2021**, *80*, 105826. [CrossRef]
34. Falsafi, S.R.; Maghsoudlou, Y.; Rostamabadi, H.; Rostamabadi, M.M.; Hamedi, H.; Hosseini, S.M.H. Preparation of Physically Modified Oat Starch with Different Sonication Treatments. *Food Hydrocoll.* **2019**, *89*, 311–320. [CrossRef]
35. Gani, A.; Gazanfar, T.; Jan, R.; Wani, S.M.; Masoodi, F.A. Effect of Gamma Irradiation on the Physicochemical and Morphological Properties of Starch Extracted from Lotus Stem Harvested from Dal Lake of Jammu and Kashmir, India. *J. Saudi Soc. Agric. Sci.* **2013**, *12*, 109–115. [CrossRef]
36. Nadir, A.S.; Helmy, I.; Nahed, M.; Abdelmaguid; Waafa, M.; Abozeid, M.; Ramadan, M.T. Modification of Potato Starch by Some Different Physical Methods and Utilization in Cookies Production. *Int. J. Curr. Microbiol. Appl. Sci.* **2015**, *4*, 556–569.
37. Singh, R.; Sharanagat, V.S. Physico-Functional and Structural Characterization of Ultrasonic-Assisted Chemically Modified Elephant Foot Yam Starch. *Int. J. Biol. Macromol.* **2020**, *164*, 1061–1069. [CrossRef]
38. Wang, H.; Xu, K.; Ma, Y.; Liang, Y.; Zhang, H.; Chen, L. Impact of Ultrasonication on the Aggregation Structure and Physicochemical Characteristics of Sweet Potato Starch. *Ultrason. Sonochem.* **2020**, *63*, 104868. [CrossRef] [PubMed]
39. Bashir, K.; Aggarwal, M. Physicochemical, Thermal and Functional Properties of Gamma Irradiated Chickpea Starch. *Int. J. Biol. Macromol.* **2017**, *97*, 426–433. [CrossRef]
40. Majeed, T.; Wani, I.A.; Hussain, P.R. Effect of Dual Modification of Sonication and γ-Irradiation on Physicochemical and Functional Properties of Lentil (*Lens culinaris* L.) Starch. *Int. J. Biol. Macromol.* **2017**, *101*, 358–365. [CrossRef]
41. Palav, T. Impact of Microwave Heating on the Physico-Chemical Properties of a Starch-Water Model System. *Carbohydr. Polym.* **2007**, *67*, 596–604. [CrossRef]
42. Nawaz, H.; Shad, M.A.; Saleem, S.; Khan, M.U.A.; Nishan, U.; Rasheed, T.; Bilal, M.; Iqbal, H.M.N. Characteristics of Starch Isolated from Microwave Heat Treated Lotus (*Nelumbo nucifera*) Seed Flour. *Int. J. Biol. Macromol.* **2018**, *113*, 219–226. [CrossRef] [PubMed]
43. Zeng, S.; Chen, B.; Zeng, H.; Guo, Z.; Lu, X.; Zhang, Y.; Zheng, B. Effect of Microwave Irradiation on the Physicochemical and Digestive Properties of Lotus Seed Starch. *J. Agric. Food Chem.* **2016**, *64*, 2442–2449. [CrossRef] [PubMed]
44. Dundar, A.N.; Gocmen, D. Effects of Autoclaving Temperature and Storing Time on Resistant Starch Formation and Its Functional and Physicochemical Properties. *Carbohydr. Polym.* **2013**, *97*, 764–771. [CrossRef]
45. Tung, C.-Y.M.; Dynes, P.J. Relationship between Viscoelastic Properties and Gelation in Thermosetting Systems. *J. Appl. Polym. Sci.* **1982**, *27*, 569–574. [CrossRef]
46. Rocha, T.S.; Demiate, I.M.; Franco, C.M.L. Características estruturais e físico-químicas de amidos de mandioquinha-salsa (*Arracacia xanthorrhiza*). *Food Sci. Technol.* **2008**, *28*, 620–628. [CrossRef]
47. Albano, K.M.; Franco, C.M.L.; Telis, V.R.N. Rheological Behavior of Peruvian Carrot Starch Gels as Affected by Temperature and Concentration. *Food Hydrocoll.* **2014**, *40*, 30–43. [CrossRef]
48. Chen, X.; Liu, Y.; Xu, Z.; Zhang, C.; Liu, X.; Sui, Z.; Corke, H. Microwave Irradiation Alters the Rheological Properties and Molecular Structure of Hull-Less Barley Starch. *Food Hydrocoll.* **2021**, *120*, 106821. [CrossRef]
49. Zhong, Y.; Tian, Y.; Liu, X.; Ding, L.; Kirkensgaard, J.J.K.; Hebelstrup, K.; Putaux, J.L.; Blennow, A. Influence of Microwave Treatment on the Structure and Functionality of Pure Amylose and Amylopectin Systems. *Food Hydrocoll.* **2021**, *119*, 106856. [CrossRef]
50. Wang, S.; Hu, X.; Wang, Z.; Bao, Q.; Zhou, B.; Li, T.; Li, S. Preparation and Characterization of Highly Lipophilic Modified Potato Starch by Ultrasound and Freeze-Thaw Treatments. *Ultrason. Sonochem.* **2020**, *64*, 105054. [CrossRef]
51. Czechowska-Biskup, R.; Rokita, B.; Lotfy, S.; Ulanski, P.; Rosiak, J.M. Degradation of Chitosan and Starch by 360-KHz Ultrasound. *Carbohydr. Polym.* **2005**, *60*, 175–184. [CrossRef]
52. Wang, M.; Wu, Y.; Liu, Y.; Ouyang, J. Effect of Ultrasonic and Microwave Dual-Treatment on the Physicochemical Properties of Chestnut Starch. *Polymers* **2020**, *12*, 1718. [CrossRef]
53. Zhao, B.; Sun, S.; Lin, H.; Chen, L.; Qin, S.; Wu, W.; Zheng, B.; Guo, Z. Physicochemical Properties and Digestion of the Lotus Seed Starch-Green Tea Polyphenol Complex under Ultrasound-Microwave Synergistic Interaction. *Ultrason. Sonochem.* **2019**, *52*, 50–61. [CrossRef]
54. Luo, Z.; Fu, X.; He, X.; Luo, F.; Gao, Q.; Yu, S. Effect of Ultrasonic Treatment on the Physicochemical Properties of Maize Starches Differing in Amylose Content. *Starch—Stärke* **2008**, *60*, 646–653. [CrossRef]
55. Yılmaz, A.; Tugrul, N. Effect of Ultrasound-Microwave and Microwave-Ultrasound Treatment on Physicochemical Properties of Corn Starch. *Ultrason. Sonochem.* **2023**, *98*, 106516. [CrossRef]

56. Zhou, Y.; Wang, M.; Li, W.; Liu, G.; Wang, W.; Zhi, W.; Wang, M.; Wang, R.; Hu, A.; Zheng, J. Effects of Dual Modification of Lysine and Microwave on Corn Starch: In Vitro Digestibility and Physicochemical Properties. *Int. J. Biol. Macromol.* **2022**, *220*, 426–434. [CrossRef] [PubMed]
57. Mieles-Gómez, L.; Quintana, S.E.; García-Zapateiro, L.A. Ultrasound-Assisted Extraction of Mango (*Mangifera indica*) Kernel Starch: Chemical, Techno-Functional, and Pasting Properties. *Gels* **2023**, *9*, 136. [CrossRef]
58. Ramírez-Brewer, D.; Quintana, S.E.; García-Zapateiro, L.A. Effect of Microwave Treatment on Technological, Physicochemical, Rheological and Microstructural Properties of Mango (*Mangifera indica*) Kernel Starch Variety Tommy and Sugar. *LWT* **2023**, *187*, 115311. [CrossRef]
59. Morrison, W.R.; Laignelet, B. An Improved Colorimetric Procedure for Determining Apparent and Total Amylose in Cereal and Other Starches. *J. Cereal Sci.* **1983**, *1*, 9–20. [CrossRef]
60. Singleton, V.L.; Orthofer, R.; Lamuela-Raventós, R.M.B.T. Analysis of Total Phenols and Other Oxidation Substrates and Antioxidants by Means of Folin-Ciocalteu Reagent. In *Oxidants and Antioxidants Part A*; Academic Press: Cambridge, MA, USA, 1999; Volume 299, pp. 152–178. ISBN 0076-6879.
61. Re, R.; Pellegrini, N.; Proteggente, A.; Pannala, A.; Yang, M.; Rice-Evans, C. Antioxidant Activity Applying an Improved ABTS Radical Cation Decolorization Assay. *Free Radic. Biol. Med.* **1999**, *26*, 1231–1237. [CrossRef] [PubMed]
62. Yamazaki, W.T. An Alkaline Water Retention Capacity Test for the Evaluation of Cookie Baking Potentialities of Soft Winter Wheat Flours. *Cereal Chem.* **1953**, *30*, 242–246.
63. Medcalf, D.G. Wheat Starches I. Comparison of Physicochemical Properties. *Cereal Chem.* **1965**, *42*, 558–568.
64. Jiang, Q.; Gao, W.; Li, X.; Man, S.; Shi, Y.; Yang, Y.; Huang, L.; Liu, C. Comparative Susceptibilities to Alkali-Treatment of A-, B- and C-Type Starches of *Dioscorea Zingiberensis, Dioscorea Persimilis* and *Dioscorea Opposita*. *Food Hydrocoll.* **2014**, *39*, 286–294. [CrossRef]
65. López-Barraza, D.; Ortega-Ramos, A.; Torregroza-Fuentes, E.; Quintana, S.E.; García-Zapateiro, L.A. Rheological and Functional Properties of Hydrocolloids from *Pereskia Bleo* Leaves. *Fluids* **2021**, *6*, 349. [CrossRef]

gels

MDPI

Article

Self-Reactive Carbon Dioxide Absorbent with Sodium Carbonate-Based Hydrogel

Jae Young Kim and Youn Suk Lee *

Department of Packaging, Yonsei University, Wonju 26493, Republic of Korea; jerrykjy@naver.com
* Correspondence: leeyouns@yonsei.ac.kr

Abstract: Sodium carbonate is an abundant, low-cost, and low-hazard raw material widely used as a food additive and CO_2 absorbent in the food industry. However, its application in food packaging is limited because it is used in solid form, either in sachets or as a compounding ingredient in plastics. Solid sodium carbonate requires an external moisture supply for CO_2 absorption, with its performance dependent on moisture availability. This limitation hinders its commercialization in food packaging applications. We developed a sodium carbonate-based, self-reactive CO_2 absorbent hydrogel incorporating polyacrylic acid sodium salt (PAAS). This sodium carbonate hydrogel (SCH-PAAS) exhibits self-reactivity, eliminating the need for external moisture, and demonstrates a high CO_2 absorption capacity. PAAS incorporation facilitates the formation of a porous structure during gel solidification through reactions with CO_2. Increased PAAS content accelerates CO_2 absorption rates, particularly under low-temperature conditions (10 °C and 25 °C). Notably, absorption was faster at 10 °C than at 25 °C. The proposed SCH-PAAS exhibits a significantly enhanced absorption performance at low temperatures compared to conventional sodium carbonate-based materials, which exhibit reduced efficiency under such conditions. The increased gas–liquid contact area in SCH-PAAS makes it highly suitable for fresh food packaging applications, particularly under low temperatures.

Keywords: sodium carbonate-based hydrogel; self-reactive CO_2 absorbent; faster CO_2 capture at low temperature

1. Introduction

Sodium carbonate (Na_2CO_3) reacts with CO_2 and moisture to form sodium bicarbonate, as shown in Equation (1) [1]:

$$Na_2CO_3\ (s) + CO_2\ (g) + H_2O\ (g) \rightarrow 2NaHCO_3\ (s) \quad (1)$$

Various applications have been developed for CO_2 absorbents in coal-fired power plants owing to their environmental friendliness, low cost, and easy regeneration of sodium carbonate. They are effective in overcoming drawbacks such as the corrosivity and high regeneration energy of commonly used amine-based absorbents. In recent years, more effective methods have been developed, including an encapsulated Na_2CO_3 aqueous sorbent, a Na_2CO_3–carbon nanocomposite [2,3].

In the food industry, sodium carbonate is used as a CO_2 scavenger in active packaging. CO_2 increases the storage of some foods, whereas in foods such as coffee and kimchi, it causes expansion during distribution [4]. However, the low capacity of CO_2 absorption of

https://doi.org/10.3390/gels11010078

sodium carbonate compared with that of calcium hydroxide ($Ca(OH)_2$) makes it unsuitable for field applications. Calcium hydroxide reacts with CO_2, as shown in Equation (2)

$$Ca(OH)_2\ (s) + CO_2\ (g) \rightarrow CaCO_3\ (s) + H_2O\ (l) \quad (2)$$

From (1) and (2), the mass-based CO_2 absorption capacities of sodium carbonate and calcium hydroxide are 9.43×10^{-3} mol g^{-1} and 1.35×10^{-2} mol g^{-1}, respectively [5].

Moreover, the suitable temperature range for the CO_2 capture of support-free hydrated sodium carbonate powders is 30–50 °C, owing to the low efficiency caused by the synthesis of $Na_2CO_3{\cdot}7H_2O$ at lower temperatures [6]. Despite several drawbacks, using Na_2CO_3 as a CO_2-absorbent material has several advantages. First, because water is required for chemical reactions, the absorption rate can be controlled by adjusting the moisture content [7]. Second, because the absorption reaction occurs even in aqueous solutions, the concentration can be optimized and the material can be applied in various forms (such as beads, slurries, and capsules) [8–12]. Finally, the reaction product, $NaHCO_3$, is a natural deodorant that removes off-odors from food [13]. To enhance its application in food packaging, its lower mass-based CO_2 absorption capacity at lower temperatures should be improved, especially when compared to the competitors described above. Additionally, its passive–active properties, which rely entirely on the moisture supplied by food, need to be enhanced.

The key to addressing the limitations of Na_2CO_3 as a CO_2 absorbent lies in its interaction with water molecules, particularly those surrounding it. Yuanhao Cai et al. [6] demonstrated that free-flowing hydrated sodium carbonate powders containing 30 wt% water (HSCP-30) achieved a high CO_2 sorption capacity of 282 mg/g within 60 min and a rapid CO_2 uptake, with 90% saturation reached in 16 min. They attributed this performance to the role of additional water in enhancing CO_2 sorption. Similarly, Xiaoyang Shi et al. [14] reported a reversible chemical reaction in Na_2CO_3 nanopores driven by water content. Reducing the water molecules in these pores promoted the hydrolysis of $CO_3{}^{2-}$ to HCO^{3-} and OH^-, enabling the development of a nano-structured CO_2 sorbent that spontaneously binds CO_2 from ambient air.

One effective method for controlling water content is using superabsorbent polymers, which can be converted into hydrogels. These hydrogels, composed of cross-linked networks of hydrophilic polymers, can absorb up to 400 times their weight in deionized water or 150–300 times their weight in irrigation water. Their cross-linked structure enables the swelling of the three-dimensional network without polymer dissolution [15,16]. Considering these properties, superabsorbent hydrogels are widely used in sanitary products, food packaging materials, water-retaining aids in agriculture, and pharmaceutical drug delivery systems [17].

I have developed a sodium carbonate-based hydrogel (SCH) using superabsorbent polymers to enhance Na_2CO_3's application as a CO_2 absorbent in food packaging materials. The SCH exhibits self-reactivity and a high CO_2 absorption capacity, eliminating the need for an external water supply and significantly improving CO_2 absorption efficiency.

2. Results and Discussion

2.1. CO_2 *Absorption Capacity*

2.1.1. Preparation of Hydrogels

At a temperature of 10 °C, the samples with varying concentrations of polyacrylic acid sodium salt (PAAS) (denoted as SCH-PAAS/X, where X represents the symbolic mass percentage of PAAS in the mixture) to evaluate their CO_2 absorption capacity are shown in Table 1. The mixing ratio of sodium carbonate to water was fixed at 40:60 wt%, and

PAAS was added at concentrations of 0, 0.25, 0.5, 0.75, and 1.0 wt%. Glycerin was added at a concentration of 9.0 wt% as a viscosity modifier, and phenolphthalein was used as a pH indicator.

Table 1. Sodium carbonate-based hydrogel (SCH) composition with PAAS.

Sample	Materials (wt%)			
	Sodium Carbonate	Water	PAAS	Glycerin
SCH-PAAS/0	36.40	54.60	0.00	9.00
SCH-PAAS/25	36.30	54.45	0.25	
SCH-PAAS/50	36.20	54.30	0.50	
SCH-PAAS/75	36.10	54.15	0.75	
SCH-PAAS/100	36.00	54.00	1.00	

Hydrogel with three droplets of phenolphthalein as pH indicator.

2.1.2. Determination of CO_2 Absorption Capacity

To evaluate the CO_2 absorption capacity of the SCHs, a pressure drop measurement method was employed in a closed chamber corresponding to the absorption of CO_2 by the material. The results were expressed as carbonation reaction capacity. The carbonation reaction capacity (R_C, mg CO_2/g Na_2CO_3), defined as the amount of CO_2 absorbed per gram of sodium carbonate, was used to compare the absorption capacities of the hydrogels on an equal basis, as the quantity of loaded Na_2CO_3 varied among samples.

In all the experimental groups where PAAS was added (Figure 1), CO_2 absorption capacity significantly improved compared to the control groups, which are without the addition of PAAS at 10 °C. From Equation (1), the theoretical R_C value of sodium carbonate was calculated to be 415.23 mg CO_2/g Na_2CO_3. The larger the amount of PAAS, the faster it took to reach the 100% theoretical R_C value at low temperatures. The R_Cs of SCH-PAAS/50, 75, and 100 were 429.34, 433.90, and 410.73 mg CO_2/g Na_2CO_3 after 6 h, respectively. There was no significant difference between SCH-PASS/50, 70, and 100. After 6 h of reaction with CO_2 samples containing more than 0.5 wt%, PAAS exhibited absorption rate was 1.8 times faster than the control and more than twice as fast compared to the 40% sodium carbonate solution.

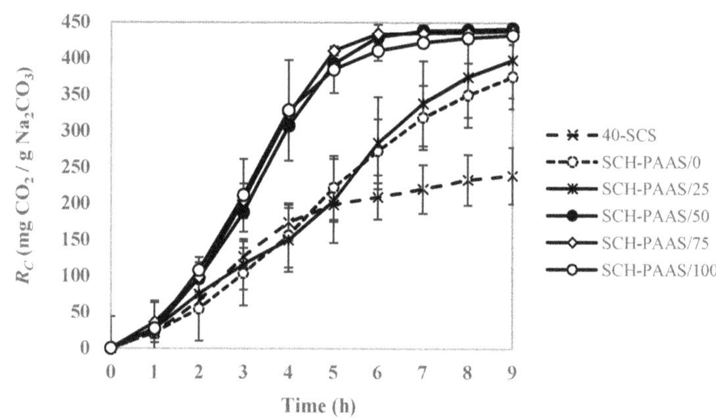

Figure 1. Changes in carbonation reaction capacity (R_C) in different SCH-PAASs as a function of time at 10 °C.

To examine the effect of temperature on the CO_2 absorption functionality of hydrogel materials, the absorption performance was evaluated at 25 °C. The R_C of SCH-PAAS/50 was 416.17 mg CO_2/g Na_2CO_3 after 9 h in the chamber, reaching 100% of the theoretical carbonation capacity and making it the fastest among the experimental groups, as shown in Figure 2. This performance demonstrated a CO_2 absorption capacity four times higher than that of the sample without PAAS (SCH-PAAS/0) after 9 h.

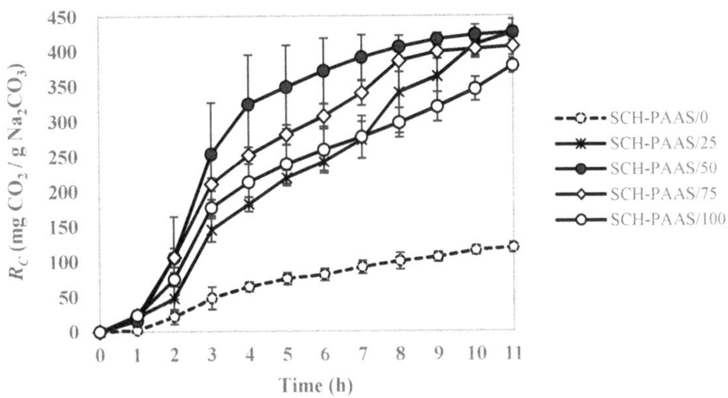

Figure 2. Changes in carbonation reaction capacity (R_C) in different SCH-PAASs as a function of time at 25 °C.

The result at a temperature of 25 °C indicates that the CO_2 absorption capacity increased with the PAAS content up to 0.5 wt%. However, a further increase in PAAS content resulted in a decrease in the CO_2 absorption capacity. This may be due to the increased gel density with higher PAAS content and the pore-clogging phenomenon caused by the narrowing of pores from the reaction-produced $NaHCO_3$ during the reaction with CO_2 [12].

Based on the experimental results at 25 °C, shown in Table 2, a comparison of the total CO_2 absorption capacity (A_C, mg CO_2/g sorbent), defined as the amount of CO_2 absorbed per 1 g sorbent, revealed that after 9 h, the sample with 0.5 wt% PAAS content absorbed 150.66 mg CO_2/g sorbent. Samples with 0.25 wt% and 0.75 wt% PAAS content absorbed 132.01 mg CO_2 and 143.84 mg CO_2/g sorbent, respectively. Therefore, based on the experiments conducted at 10 °C and 25 °C, the optimal PAAS content for achieving the most effective CO_2 absorption capacity is evaluated to be 0.5 wt%.

Table 2. CO_2 absorption capacities of the hydrogels after a 9 h reaction in the chamber at 25 °C.

Sample	A_C * mg CO_2/g Sorbent	R_C mg CO_2/g Na_2CO_3	
		Experimental	Theoretical
SCH-PAAS/0	38.51 ± 2.44 [a]	105.80 ± 6.71 [a]	
SCH-PAAS/25	132.01 ± 20.99 [b,c]	363.67 ± 57.81 [b,c]	
SCH-PAAS/50	150.66 ± 3.16 [c]	416.17 ± 8.75 [c]	415.23
SCH-PAAS/75	143.84 ± 4.14 [c]	398.45 ± 11.47 [c]	
SCH-PAAS/100	115.27 ± 7.49 [b]	320.18 ± 20.82 [b]	

* Values followed by the same letter (a, b, c) are not significantly different ($p < 0.05$) according to Duncan's multiple range test.

As mentioned above, the reaction time was faster at 10 °C compared to 25 °C because of the increase in CO_2 solubility with decreasing temperatures. This is attributed to the solubility of CO_2 in water increasing from 0.149 g CO_2/100 g water at 25 °C to 0.234 g CO_2/100 g water at 10 °C [18,19]. The increase in carbon dioxide solubility enhances the concentration of H^+ and HCO_3^- on the surface Na_2CO_3. The kinetics of CO_2 captured on Na_2CO_3 is not controlled by the surface carbonation process but is diffusion-controlled [20,21]. According to Fick's first law, as shown in the equation below, the diffusion flux (*J*) is proportional to the concentration (C). Therefore, the increased carbon dioxide solubility at 10 °C compared to 25 °C is considered to enhance the carbon dioxide absorption rate of the hydrogel.

$$J = -D\frac{\partial C}{\partial x}$$

where J is the diffusion flux (mol/m^2·s), D is the diffusion coefficient (m^2/s), C is the concentration, and χ is the distance.

Its absorption capacity demonstrates a high performance. The observed CO_2 absorption capacities of the composite sorbents containing Na_2CO_3 were primarily determined by their Na_2CO_3 loadings. Theoretically, all the Na_2CO_3 on the surface of the sorbents can be used for CO_2 absorption according to Equation (1) in a humid atmosphere. However, in many studies, the theoretical absorption capacities were not achievable because the active ingredients could not be distributed equally on the surface of the sorbents [11]. This study addresses the issue by demonstrating that PAAS, functioning as a water-lock material, can supply water to Na_2CO_3 molecules throughout the gel.

2.2. *Properties of Hydrogel with CO_2 Absorption*

2.2.1. Phase Change

The carbonation reaction described by Equation (3) consumes hydroxide ions (OH^-), leading to a decrease in the pH of the hydrogel from the CO_2 absorption.

$$OH^- + CO_2 \rightarrow HCO_3^- \quad (3)$$

To visually assess CO_2 absorption, we used phenolphthalein as a pH indicator in the preparation of the hydrogel. The pH of the hydrogel before reacting with CO_2 was measured at 11 (Milwaukee MW102 PRO + 2-in-1), indicating a pink color. After 6 h of reaction with CO_2, the pH of the hydrogel was neutralized, causing it to turn white because of the colorless nature of the indicator at a neutral pH (Figure 3).

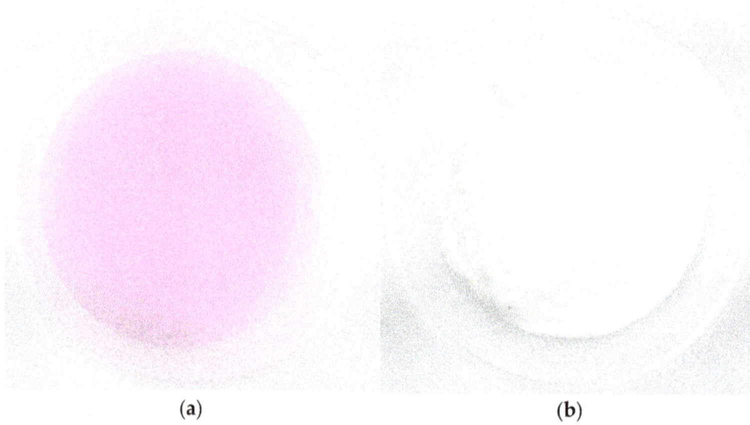

Figure 3. Change in SCH-PAAS/50 color after the reaction with CO_2 for 6 h at 10 °C: (**a**) before the reaction with CO_2, (**b**) after the reaction with CO_2.

2.2.2. Chemical Change

Figure 4 shows the FI-IR spectra of pure Na_2CO_3, before and after its reaction with the CO_2 of SCH-PAAS/0 and 50. Joshi et al. [22] used the independent peaks of each substance to quantify Na_2CO_3 and $NaHCO_3$ in a solid mixture: Na_2CO_3 at 1450 cm^{-1}, $NaHCO_3$ at 1000 and 1923 cm^{-1}. All the samples before the reaction exhibit a strong peak characteristic at 1450 cm^{-1} (Na_2CO_3) in Figure 5. After the reaction with CO_2 for 6 h, SCH-PAAS/50 showed two peaks at 1000 and 1923 cm^{-1} ($NaHCO_3$) and a significantly decreased peak at 1450 cm^{-1}, whereas the peak at 1450 cm^{-1} remained in SCH-PAAS/0, and the two characteristic peaks of $NaHCO_3$ did not appear. This indicates that the conversion of $NaHCO_3$ from Na_2CO_3 in SCH-PAAS/50 was highly effective.

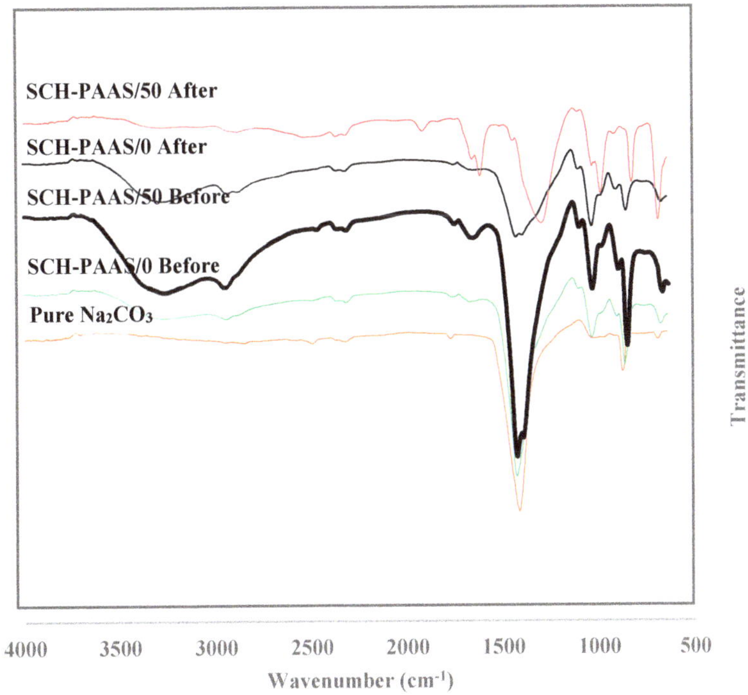

Figure 4. FT−IR of SCH−PAASs before and after the reaction with CO_2.

Figure 5 shows peaks at 906, 860, and 679 cm^{-1}, which correspond to the monohydrate [23] in SCH-PAAS/50 before the reaction. However, these peaks were not observed after the reaction with CO_2. During the preparation of the hydrogel, the conditions necessary for the formation of the monohydrate, as described in Equation (4), are as follows: excess water vapor, a low temperature (below 333 K), and an insufficiently high concentration of CO_2 in the gas phase [24]

$$Na_2CO_3\ (s) + H_2O\ (g) \leftrightarrow Na_2CO_3{\cdot}H_2O\ (s) \quad (4)$$

$$Na_2CO_3{\cdot}H_2O\ (s) + CO_2\ (g) \rightarrow 2NaHCO_3\ (s) \quad (5)$$

Thus, when the hydrogel absorbs CO_2, the reactions shown in Equation (5) are also observed. The presence of this monohydrate may explain the result of rapid CO_2 absorption at 10 °C compared to 25 °C, because of the increased solubility of CO_2 as the temperature decreases.

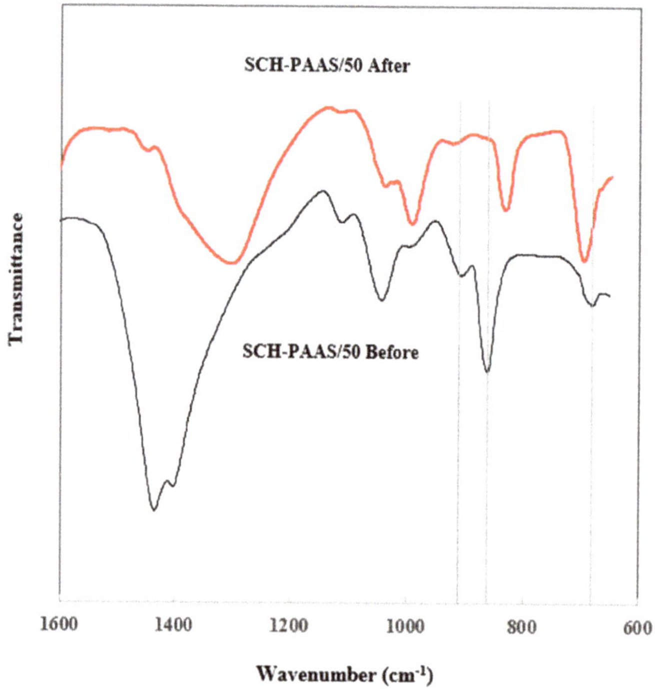

Figure 5. FT-IR of SCH-PAAS/50 before and after the reaction with CO_2.

2.2.3. Structure Change

In the 10 °C experiment, samples with more than 0.5 wt% PAAS exhibited the same absorption capacity. However, in the 25 °C experiment, a difference in CO_2 absorption performance based on the PAAS content was observed. To examine the morphological differences in CO_2 absorption due to PAAS, a SEM analysis was conducted on the samples with 0 wt% PAAS and the sample with 0.5 wt% PAAS, which showed the fastest CO_2 absorption at 25 °C.

Numerous network-shaped pores were observed on the surface of the SCH-PAAS/50 sample containing 0.5 wt% polyacrylic acid salt (Figure 6). After the reaction with CO_2, numerous porous structures were formed, although the pore sizes decreased to 3.6~5.2 μm compared to those before the reaction. Conversely, on the surface of the SCH-PAAS/0 sample, small and dense Na_2CO_3 particles were observed before the reaction. However, after the reaction with CO_2, large elongated rod-like and plate-shaped $NaHCO_3$ solid particles were observed. Luo et al. [24] reported that the macropores on the surface of Na_2CO_3 nearly disappeared within 10 min, depending mainly on the higher CO_2 and H_2O concentrations, which induced the bicarbonate formation of Na_2CO_3.

Therefore, the well-developed network structure of SCH-PAAS/50 was interpreted as having an effect of increasing the gas–liquid contact area.

The addition of PAAS in the preparation of sodium bicarbonate-based hydrogels affects the formation of the gel's porous structure, particularly enabling the formation of a greater number of porous structures when the gel solidifies through the reaction with CO_2. This effect occurs because of the uniform and widespread distribution of water molecules and the monohydrate within the cross-linked network structure of PAAS, followed by the conversion of Na_2CO_3 to $NaHCO_3$ through the carbonation reaction.

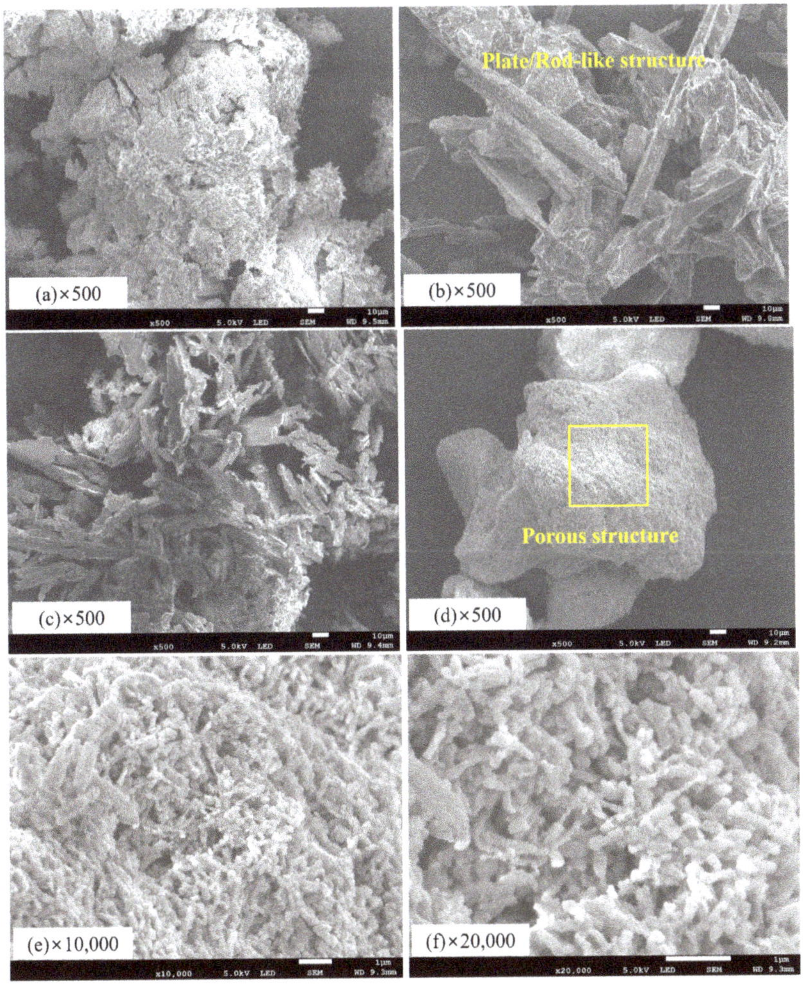

Figure 6. SEM images of SCH with different PAAS ratios: (**a**) before and (**b**) after the reaction with the CO_2 of SCH-PAAS/0; (**c**) before and (**d**) after the reaction with the CO_2 of SCH-PAAS/50; (**e**,**f**) enlargement of the special section of (**d**).

3. Conclusions

A hydrogel with CO_2 absorption functionality was developed by adding the superabsorbent polymer PAAS to a sodium carbonate solution. The addition of PAAS overcomes the limitations of traditional sodium carbonate-based CO_2 absorbent materials, which require an external water supply for activation. As the hydrogel inherently contains water, it reacts immediately with CO_2, and the inclusion of PAAS enhances its CO_2 absorption capacity.

Water and sodium carbonate distributed along the PAAS polymer chains prevented the aggregation of solid sodium bicarbonate particles formed as a result of CO_2 absorption and contributed to the formation of a porous network structure. Owing to this porous structure, CO_2 was smoothly supplied to the hydrogel, increasing the gas–liquid contact area and resulting in rapid CO_2 absorption efficiency. This increase in the gas–liquid contact area, combined with the enhanced solubility of CO_2 as the temperature decreased from

25 °C to 10 °C, produced a synergistic effect, resulting in faster CO_2 absorption at lower temperatures. Finally, a highly efficient, self-reactive CO_2 absorbent was developed.

4. Materials and Methods

4.1. Materials

Sodium carbonate (99.0%, Samchun Chemicals, Pyeongtaek-si, Republic of Korea), polyacrylic acid salt (Sodium polyacrylate, polymerization degree 22,000~70,000, FUJIFILM Wako Pure Chemical Corporation, Osaka, Japan), and glycerin (99.0%, Samchun Chemicals, Pyeongtaek-si, Republic of Korea) were used as reactants. A phenolphthalein indicator was used to visually determine the decrease in pH owing to CO_2 absorption. Deionized water was used in all the fabrication procedures.

4.2. Preparation of the Hydrogels

The mixing ratio of sodium carbonate to water was fixed at 40:60 wt%, and polyacrylic acid sodium salt (PAAS) was added at concentrations of 0, 0.25, 0.5, 0.75, and 1.0 wt%. As shown in Figure 7, the mixture was stirred using an overhead stirrer with a 4-blade impeller for 5 min at 25 °C. Subsequently, 9 wt% glycerin and 3 drops of phenolphthalein indicator were added, and the mixture was stirred for 40 min at 25 °C. Following that, it was cooled in a water bath at 10 °C, while stirring it with a spatula, for 10 min. The hydrogel was stored in bottles in a 25 °C incubator and subsequently used for the experiments.

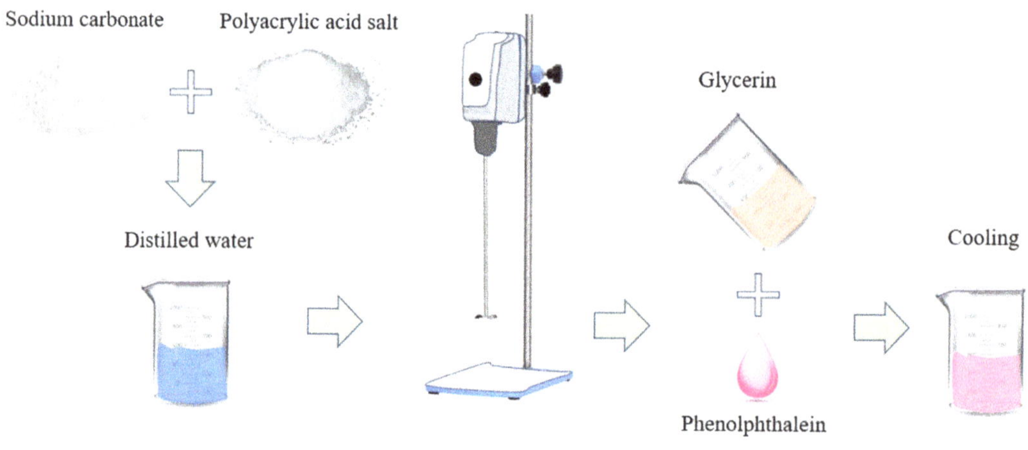

Figure 7. Schematic diagram of the sodium carbonate-based hydrogel preparation.

4.3. Properties of the Hydrogels

The morphology of the hydrogels was analyzed using a JSM-7001F (JEOL Ltd., Tokyo, Japan) field-emission scanning electron microscope (FE-SEM). To characterize the chemical structure, all spectra were recorded from 4000 to 650 cm^{-1} using the FT-IR spectrometer (FT/IR-4100, Jasco, Tokyo, Japan), and all measurements were made using 32 scans at 4 cm^{-1} resolution. The samples used for the Fourier-transform infrared (FT-IR) analysis and FE-SEM measurements were freeze-dried for 24 h before use.

4.4. CO_2 Absorption Capacity

The evaluation of CO_2 absorption capacity followed the pressure drop method of Vericella et al. [2], which was modified to align with the hydrogel experiment (Figure 8). Initially, a chamber with 5 g of hydrogel was evacuated to −60 kPa using a vacuum pump and subsequently filled with pure CO_2 gas. The pressure within the chamber was monitored as a function of time using a pressure gauge at 10, 25 °C. In each experiment, a blank chamber without hydrogel was included to monitor the absorption of CO_2 by residual ambient moisture. The hydrogel used in the absorption experiment was molded with a diameter of 45.0 mm and a thickness of 3.0 mm.

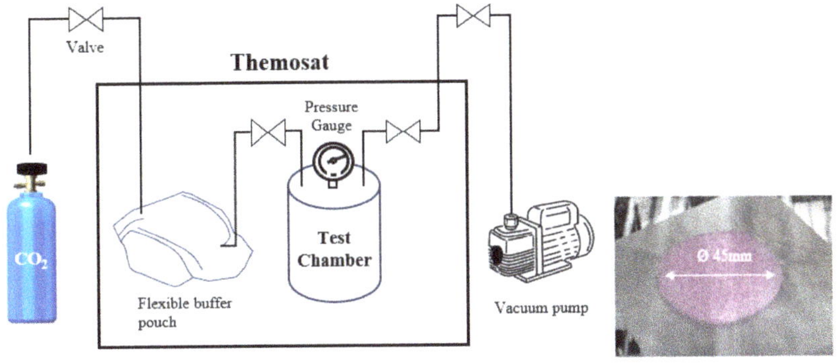

Figure 8. Schematic of pressure drop measurement method to evaluate CO_2 absorption capacity.

The CO_2 absorption capacity was calculated using the methods described by Wang et al. [25] and Zhao et al. [10]. From the Ideal Gas Law, the amount of CO_2 absorption (n_A, mol) in a chamber can be calculated from the pressure drop as follows:

$$n_A = \frac{P_g \cdot V_C}{R \cdot T}$$

where V_c is the volume of the sample chamber (L), T is absolute temperature (K), R is the gas constant (0.08205 L·atm·K^{-1}·mol^{-1}), and P_g is the pressure (atm).

The CO_2 absorption capacity of the hydrogel is expressed as the total CO_2 absorption capacity (A_C, mg CO_2/g sorbent), which is defined as the amount of CO_2 absorbed per 1 g sorbent, and the carbonation reaction capacity (R_C, mg CO_2/g Na_2CO_3), which is defined as the amount of CO_2 absorbed per 1 g alkali metal carbonate.

$$A_C = \frac{1000 \cdot n_A \cdot M_{CO2}}{m}$$

$$R_C = \frac{A_C}{\alpha}$$

where n_A (mol) is the total amount of CO_2 absorbed in carbonation (mol), M_{CO2} is the molecular mass of CO_2, and m and α are the mass of hydrogel and the loaded amount of Na_2CO_3 in the sample, respectively.

4.5. Statistics Analysis

The data analysis for the experimental results was performed using SPSS software (IBM SPSS Statistics 27, SPSS Inc. Chicago, IL, USA). The data were subjected to an analysis of variance (ANOVA). The statistical significance of differences between the mean values was established at $p < 0.05$, and Duncan's multiple range test was used for all statistical analyses.

Author Contributions: Conceptualization, J.Y.K. and Y.S.L.; methodology, J.Y.K.; validation, J.Y.K.; formal analysis, J.Y.K.; investigation, J.Y.K.; writing—original draft preparation, J.Y.K.; writing—review and editing, Y.S.L.; visualization, J.Y.K.; supervision, Y.S.L.; All authors have read and agreed to the published version of the manuscript.

Funding: This research was supported by the Korean Institute of Planning and Evaluation for Technology in Food, Agriculture, and Forestry (IPET) through the High Value-Added Food Technology Development Program, funded by the Ministry of Agriculture, Food, and Rural Affairs (MAFRA) 123001-02.

Institutional Review Board Statement: Not applicable.

Informed Consent Statement: Not applicable.

Data Availability Statement: The original contributions of this study are included in the article. Further inquiries can be directed to the corresponding author.

Conflicts of Interest: The authors declare that the research was conducted in the absence of any commercial or financial relationships that could be construed as potential conflicts of interest.

References

1. Liang, Y.; Harrison, D.P. Carbon dioxide capture using dry sodium-based sorbents. *Energy Fuels* **2004**, *18*, 569–575. [CrossRef]
2. Vericella, J.J.; Baker, S.E.; Stolaroff, J.K.; Duoss, E.B.; Hardin, J.O.; Lewicki, J.; Glogowski, E.; Floyd, W.C.; Valdez, C.A.; Smith, W.L.; et al. Encapsulated liquid sorbents for carbon dioxide capture. *Nat. Commun.* **2015**, *6*, 6124. [CrossRef]
3. Nasiman, T.; Kanoh, H. Preparation of the Na_2CO_3-Carbon nanocomposite and its CO_2 capture. *Energy Fuels* **2018**, *32*, 12689–12694. [CrossRef]
4. Lee, H.-G.; Jeong, S.; Yoo, S. Development of food packaging materials containing calcium hydroxide and porous medium with carbon dioxide-adsorptive function. *Food Packag. Shelf Life* **2019**, *21*, 100352. [CrossRef]
5. Lee, D.S. Carbon dioxide absorbers for food packaging applications. *Trends Food Sci. Technol.* **2016**, *57*, 146–155. [CrossRef]
6. Cai, Y.; Wang, W.; Li, L.; Wang, Z.; Wang, S.; Ding, H.; Zhang, Z.; Sun, L.; Wang, W. Effective capture of carbon dioxide using hydrated sodium carbonate powders. *Materials* **2018**, *11*, 183. [CrossRef] [PubMed]
7. Park, S.-W.; Sung, D.-H.; Choi, B.; Oh, K.; Moon, K. Sorption of carbon dioxide onto sodium carbonate. *Sep. Sci. Technol.* **2006**, *41*, 2665–2684. [CrossRef]
8. Yu, W.; Wang, T.; Park, A.-H.A.; Fang, M. Toward sustainable energy and materials: CO_2 capture using microencapsulated sorbents. *Ind. Eng. Chem. Res.* **2020**, *59*, 9746–9759. [CrossRef]
9. Valluri, S.; Kawatra, S.K. Use of frothers to improve the absorption efficiency of dilute sodium carbonate slurry for post combustion CO_2 capture. *Fuel Process. Technol.* **2021**, *212*, 106620. [CrossRef]
10. Zhao, C.; Chen, X.; Anthony, E.J.; Jiang, X.; Duan, L.; Wu, Y.; Dong, W.; Zhao, C. Capturing CO_2 in flue gas from fossil fuel-fired powder plants using dry regenerable alkali metal-based sorbent. *Prog. Energy Combust. Sci.* **2013**, *39*, 515–534. [CrossRef]
11. Yu, F.; Wu, Y.; Zhang, W.; Cai, T.; Xu, Y.; Chen, X. A novel aerogel sodium-based sorbent for low temperature CO_2 capture. *Greenh. Gas. Sci. Technol.* **2016**, *6*, 561–573. [CrossRef]
12. Tuwati, A.; Fan, M.; Russell, A.G.; Wang, J.; Dacosta, H.F.M. New CO_2 sorbent synthesized with nanoporous $TiO(OH)_2$ and K_2CO_3. *Energy Fuels* **2013**, *27*, 7628–7636. [CrossRef]
13. Jeong, S.; Yoo, S. Preparation and properties of sodium bicarbonate-incorporated LDPE films with deodorizing function for kimchi packaging. *Packag. Technol. Sci.* **2017**, *30*, 781–790. [CrossRef]
14. Shi, X.; Xiao, H.; Lackner, K.S.; Chen, X. Capture CO_2 from ambient air using nanoconfined ion hydration. *Angew. Chem. Int. Ed.* **2016**, *55*, 4026–4029. [CrossRef]
15. Tomar, R.S.; Gupta, I.; Singhal, R.; Nagpal, A.K. Synthesis of poly (acrylamide-co-acrylic acid) based superabsorbent hydrogels: Study of network parameters and swelling behavior. *Polym. Plast. Technol. Eng.* **2007**, *465*, 481–488. [CrossRef]
16. Cannazza, G.; Cataldo, A.; De Benedetto, E.; Demitri, C.; Madaghiele, M.; Sannino, A. Experimental assessment of the use of a novel superabsorbent polymer (SAP) for the optimization of water consumption in agricultural irrigation process. *Water* **2014**, *6*, 2056–2069. [CrossRef]
17. Kim, J.S.; Kim, D.H.; Lee, Y.S. The influence of monomer composition and surface-crosslinking condition on biodegradation and gel strength of super absorbent polymer. *Polymer* **2021**, *13*, 663. [CrossRef]
18. Dodds, W.S.; Stutzman, L.F.; Sollami, B.J. Carbon dioxide solubility in water. *Ind. Eng. Chem. Chem. Eng. Data Ser.* **1956**, *1*, 92–95. [CrossRef]
19. Lee, S.-W.; Chae, S.; Bang, J.-H. Synthesis of Na compounds from sodium concentrated solution using carbonation and cryo-crystallization. *J. Korean Inst. Resour. Recycl.* **2020**, *29*, 58–66. [CrossRef]
20. Cai, T.; Johnson, J.K.; Wu, Y.; Chen, X. Toward understanding the kinetics of CO_2 capture on sodium carbonate. *ACS Appl. Mater. Interface* **2020**, *11*, 9033–9041. [CrossRef]
21. Cai, T.; Chen, X.; Johnson, J.K.; Wu, Y.; Ma, J.; Liu, D.; Liang, C. Understanding and improving the kinetics of bulk carbonation on sodium carbonate. *J. Phys. Chem. C* **2020**, *124*, 23106–23115. [CrossRef]

22. Joshi, S.; Kalyanasundaram, S.; Balasubramanian, V. Quantitative analysis of sodium carbonate and sodium bicarbonate in solid mixtures using fourier transform infrared spectroscopy. *Appl. Spectrosc.* **2013**, *67*, 841–845. [CrossRef]
23. Rodriquez-Mosqudea, R.; Bramer, E.A.; Brem, G. CO_2 capture from ambient air using hydrated Na_2CO_3 supported on activated carbon honeycombs with application to CO_2 enrichment in greenhouse. *Chem. Eng. Sci.* **2018**, *189*, 114–122. [CrossRef]
24. Luo, H.; Kanoh, H. Fundamental in CO_2 capture of Na_2CO_3 under a moist condition. *J. Energy Chem.* **2017**, *26*, 972–983. [CrossRef]
25. Wang, H.J.; Jo, Y.H.; An, D.S.; Rhim, J.-W.; Lee, D.S. Properties of agar-based CO_2 absorption film containing Na_2CO_3 as active compound. *Food Packag. Shelf Life* **2015**, *4*, 36–42. [CrossRef]

gels

Article

Development and Quality Enhancement of Fried Fish Cake Prototype with Transglutaminase, Trehalose, and Herbal Oil for Room Temperature Distribution

Ye Youl Kim [1,†], Khawaja Muhammad Imran Bashir [2,3,†], Seonyeong Kang [1], Yongjoon Park [1], Jae Hak Sohn [4,5,*] and Jae-Suk Choi [2,*]

1 Samjin Food Co., Ltd., Busan 49036, Republic of Korea; dudndi1324@samjinfood.com (Y.Y.K.); glorysun@samjinfood.com (S.K.); sj0001@samjinfood.com (Y.P.)
2 Department of Seafood Science and Technology, Institute of Marine Industry, Gyeongsang National University, Tongyeong 53064, Republic of Korea; imranbashir@gnu.ac.kr
3 German Engineering Research and Development Center for Life Science Technologies in Medicine and Environment, Busan 46742, Republic of Korea
4 Department of Food Science and Culinary Arts, College of Health and Welfare, Silla University, Busan 46958, Republic of Korea
5 Seafood Research Center, IACF, Silla University, 606, Advanced Seafood Processing Complex, Busan 49277, Republic of Korea
* Correspondence: jhsohn@silla.ac.kr (J.H.S.); jsc1008@gnu.ac.kr (J.-S.C.); Tel.: +82-51-999-5629 (J.H.S.); +82-55-772-9142 (J.-S.C.)
† These authors contributed equally to this work.

Abstract: This study focuses on developing a fried fish cake prototype with improved quality and extended shelf-life, enabling room-temperature distribution through an innovative high-temperature and high-process retort method. Surimi-based products typically necessitate cold storage and a refrigerated distribution system, affecting their physical properties and flavor while escalating costs. By incorporating Transglutaminase (TGase), trehalose, and herbal oils, and optimizing the heating process using the response surface methodology, this research addresses challenges related to changes in physical properties, color, and off-flavors during high-temperature and high-pressure treatment. The addition of 0.37% ACTIVA-K TGase significantly enhanced gel strength by promoting protein cross-linking, while 0.75% trehalose improved color stability by suppressing browning, thus enhancing visual appeal. A 0.1% concentration of bay oil effectively enhanced the flavor profile by masking undesirable odors without compromising the sensory quality. Optimized processing conditions maximized DPPH radical scavenging activity, whiteness, and gel strength, ensuring superior product quality and safety. Nutritional analysis confirmed a balanced composition of moisture, protein, essential amino acids, and minerals, in accordance with Korean national standards for acid values. Microstructural examination revealed a uniform network structure, contributing to excellent texture and sensory evaluations. Shelf-life predictions indicated a storage duration of approximately 19 months, surpassing commercially available products and offering a competitive edge. This novel approach allows surimi-based products to be stored and distributed at room temperature, while also providing the potential for increased profitability.

Keywords: gel strength; high-pressure treatment; high-temperature processing; product enhancement; response surface analysis; storage duration; surimi-derived products

1. Introduction

In the global food industry, surimi—a refined concentrate of myofibrillar proteins—is lauded for its nutritional benefits, primarily its high protein and low-fat content. Derived from fish meat through the removal of sarcoplasmic proteins, blood, fat, and pigments, surimi forms the foundation for a variety of health-oriented food products [1]. Fish cakes, among the myriads of surimi products, are particularly popular due to their versatility in preparation methods like steaming, baking, and frying. In Korea, frying is preferred for its ability to enhance the visual appeal, flavor, and texture of fish cakes, as well as its efficiency in reducing cooking time [2].

As of 2022, Korea's domestic production of fried fish cakes was approximately 100,000 tons, accounting for 56% of all surimi products. These products are primarily distributed under refrigeration, offering a shelf-life of up to 10 days for unsterilized and up to 50 days for sterilized varieties. The export volume, ranging between 9814 and 11,100 tons, is expected to grow [3]. However, frozen exports face quality challenges such as sponginess and drip loss upon thawing, which restricts market expansion both domestically and internationally.

To facilitate room temperature distribution, effective sterilization processes are crucial. Traditional methods like high temperature and high-pressure treatments are favored for their safety and convenience, yet they can lead to nutrient degradation, texture deterioration, and undesirable sensory effects such as off-flavors or discoloration [4]. This has spurred research into enhancing texture with additives like transglutaminase (TGase) [5,6], konjac [7,8], and carrageenan [9,10].

TGase is a promising agent for improving the quality of meat products by enhancing gel strength through the formation of ε-(γ-glutamyl)-lysine cross-links [11–14]. It is particularly effective in surimi products, promoting cross-linking and protein aggregation at elevated temperatures [4,13]. It has been reported to enhance the gel strength of surimi products [15]. While at low temperatures, the cross-linked protein structure inhibits hydrogen bond formation; at high temperatures, it promotes cross-linking and aggregation of surimi proteins [4,13].

The Maillard reaction leads to alterations in protein digestibility, protein functionality, and the sensory characteristics of food, which in turn affect the color of the food [16]. Therefore, it is crucial to manage Maillard reaction during food processing, particularly during production and storage, to preserve the desired quality of the food. In high-pressure processing, sugars like glucose, sucrose, and trehalose are incorporated into food products to prevent discoloration and extend shelf-life [16,17]. Trehalose is particularly favored in food processing due to its superior stabilizing properties [18–20]. Trehalose is a white, odorless powder with a sweetness approximately 45% that of sucrose. It is a non-reducing disaccharide comprising two α-glucose molecules valued for its ability to prevent starch aging and stabilize proteins and lipids [21]. It is naturally found in algae, yeast, fungi, plants, and insects [22]. Its applications span food processing, cosmetics, and pharmaceuticals, with research emphasizing its benefits in inhibiting starch aging [23,24], enhancing water holding capacity [25] and storage properties in agricultural product processing [24]. Despite its versatility, limited studies exist on its effectiveness in preventing color changes in processed fishery products like fish cakes [4]. Additionally, the incorporation of herbal oils, known for their antimicrobial and antioxidant properties, presents a novel approach to enhancing the sensory attributes and extending the shelf-life of food products.

This research explores the synergistic effects of TGase, trehalose, and herbal oils in developing a high-quality fried fish cake prototype suitable for room temperature distribution. The combination of TGase, trehalose, and herbal oils was selected based on our preliminary study of different combinations and the recent report by Lee et al. [4]. The study aims to optimize processing conditions that maintain product integrity during high-temperature and high-pressure treatment, thereby supporting the expansion of fish cake markets globally. Furthermore, the physicochemical and sensory quality characteristics of the newly developed prototypes, along with their extended shelf-life, are thoroughly evaluated to ensure a competitive edge in both domestic and international markets.

2. Results and Discussion

2.1. Addition of TGase and Sugars in Fried Fish Cake Prototype

2.1.1. Gel Strength Improving Effect of TGase

To evaluate the texture related to heating and TGase addition, the gel strength of the fried fish paste prototype was measured, and the results are shown in Table 1. The test product that was heat-treated with added TGase showed the highest gel strength value of 156.52 g × cm. TGase is an enzyme that forms an ε-(γ-Glu)Lys bond by inducing a cross-link between the γ-carboxyamide group of the glutaminyl residue and the ε-amino group of the lysine residue. When used in food, this enzyme is known to form cross-links between proteins, thereby improving gel formation, coagulation, heat stability, hydration, solubility, and adhesiveness of the food [26,27]. This could explain the higher gel strength observed in the samples containing TGase.

Table 1. Gel strength of the prototype according to the addition of TGase and the presence or absence of heating (unit: g × cm).

Category	Gel Strength (g × cm)	
	No TG Added	TG Added
No Heating	111.17 ± 4.76 ^{a}	111.10 ± 4.76 ^{a}
Heating	132.53 ± 3.87 ^{b}	156.52 ± 5.62 ^{b}

Values are means ± S.D. of three replications (n = 3). Different superscript letters (a, b) indicate significant differences among means by Duncan's test ($p < 0.05$).

Numerous studies have reported improvements in the textural properties of surimi and surimi-based products through the application of various protein additives [28–31]. In our study, we utilized ACTIVA TG-K to improve the texture of a newly developed fried fish cake prototype. This additive facilitates deamidation, amine incorporation, and protein cross-linking [32]. By forming ε-(γ-glutamyl) lysine bonds within the myosin heavy chain during the setting of salted fish paste, ACTIVA TG-K enhances texture, leading to a firmer and more cohesive consistency [33]. In line with previous research on the use of transglutaminases to enhance gel strength [20,32–35], our results showed a significant increase in gel strength in the fried fish cake prototype with 0.37% ACTIVA TG-K.

2.1.2. Gel Strength and Color Improvement Effect of Added Sugars

To examine the texture and color, the addition of TGase and the heating process (55 °C, 10 min) were applied in the same manner. When the gel strength, yellowness, and whiteness of the samples with different sugars added were measured, the sample with added trehalose showed the highest gel strength at 187.34 g × cm (Table 2). The elevated gel strength is attributed to trehalose's ability to maintain an active form, which stabilizes protein molecules [36].

Table 2. Gel strength, yellowness, and whiteness of fried fish cake samples with different added sugars.

Parameter	Sucrose	Glucose	Xylose	Trehalose
Gel strength (g × cm)	163.20 ± 3.56 ^{b}	150.33 ± 5.10 ^{a}	177.66 ± 6.62 ^{c}	187.34 ± 9.82 ^{c}
Yellowness	6.67 ± 0.10 ^{a}	13.03 ± 0.49 ^{b}	26.05 ± 0.69 ^{c}	6.53 ± 0.47 ^{a}
Whiteness	49.41 ± 0.57 ^{d}	28.57 ± 0.39 ^{b}	−31.06 ± 1.55 ^{a}	47.29 ± 0.67 ^{c}

Values are means ± S.D. of three replications (n = 3). Different superscript letters (a–d) indicate significant differences among means by Duncan's test ($p < 0.05$).

In examining the color change in the sample products based on the type of sugar, in terms of yellowness, the sample product with added xylose showed the highest value at 26.05, while the sample with added trehalose was the lowest at 6.53. Regarding whiteness, the product with added trehalose was the highest at 49.29, and the product with added xylose added was the lowest at −31.06 (Table 2).

When heat is applied to a mixture of fish cake dough, fish meat surimi, and sugar, the Maillard reaction occurs as the amino group of the amino acids combines with the carbonyl group of the reducing sugars. Xylose, a 5-carbon sugar, has a more reactive reducing group than other sugars, leading to a more pronounced yellowness [37]. Trehalose, a naturally occurring disaccharide composed of two glucose molecules linked by an α-1,1 glycosidic bond, is a non-reducing sugar known for its chemical stability and resistant to acids, alkalis, and the Maillard reaction. This makes it a popular choice in various food applications [38]. In this study, we observed that browning was minimized when heating the fish cake prototype with added trehalose. The improved gel strength in the fried fish cake prototype containing trehalose can be attributed to its capacity to stabilize protein molecules, thereby enhancing the overall gel structure in the fish cakes [21,36].

2.1.3. Flavor Improvement by Addition of Herbal Oil

Defrosted fish cake products often experience nutrient degradation, texture loss, changes in physicochemical properties, taste alterations, and the development of off-flavors or discoloration [39]. To address these issues, we enhanced the fish cakes with various herbal oils. The results for odor intensity and sensory evaluation of the test products with added tocopherol and various herbal oils are presented in Table 3. The fried fish cake with bay oil demonstrated a particularly suitable odor intensity (468.8 ± 22.1) and received higher scores in sensory evaluations (7.47 ± 0.2). Although the fried fish cake prototype supplemented with tocopherol had a slightly higher sensory evaluation score than the one with bay leaf oil and a lower odor intensity (388.6 ± 19.4), the fragrance was not preferred by sensory evaluators. Consequently, bay oil, which had a relatively lower odor intensity and high sensory evaluation scores, was identified as the most suitable herb oil for the fried fish cake prototype.

Table 3. Odor intensity and sensory scores of fried fish cake with herbal oil addition.

Sample	Odor Intensity (Level)	Sensory Score (Point)
Control group	391.3 ± 13.6 [a]	6.4 ± 0.1 [c]
Tocopherol	388.6 ± 19.4 [a]	7.6 ± 0.3 [d]
Bay leaf	468.8 ± 22.1 [b]	7.4 ± 0.2 [d]
Coriander	521.9 ± 20.5 [c]	4.8 ± 0.2 [a]
Cumin	499.3 ± 17.3 [b, c]	5.5 ± 0.4 [b]
Basil	512.9 ± 25.5 [c]	5.2 ± 0.3 [a, b]
Anisid	411.0 ± 36.1 [a]	5.6 ± 0.1 [b]

Values are means ± S.D. of three replications (n = 3). Different superscript letters (a–d) indicate significant differences among means by Duncan's test ($p < 0.05$).

2.2. Optimal Processing Conditions for Fried Fish Cake Prototype

The DPPH (Y_1), a dependent variable of the fried fish cake sample with added TGase, laurel oil, and trehalose, tended to increase as X_1 (amount of antioxidant added) moved from −1.682 to +1.682. For both X_2 (temperature conditions during high-temperature and high-pressure treatment) and X_3 (time conditions for high-temperature and high-pressure treatment), DPPH (Y_1) value showed a decreasing trend as it shifted from −1.682 to +1.682. Another dependent variable, whiteness (Y_2), did not change much as X_1 moved from −1.682 to +1.682. However, as X_2 and X_3 moved from −1.682 to +1.682, the whiteness (Y_2) of the prototype tended to decrease.

This indicates that the amount of antioxidant added does not significantly affect the whiteness of the prototype, whereas the temperature and time during the high-temperature and high-pressure treatment process have a significant impact. In the case of X_1, there was no significant change in the gel strength (Y_3) of the prototype as it moved from −1.682 to

+1.682. Conversely, for X_2 and X_3, there was a tendency for the gel strength (Y_3) of the prototype to decrease as it moved from −1.682 to +1.682. It is also believed that gel strength is not affected by antioxidants but is influenced by the conditions of high-temperature and high-pressure treatment (Figure 1).

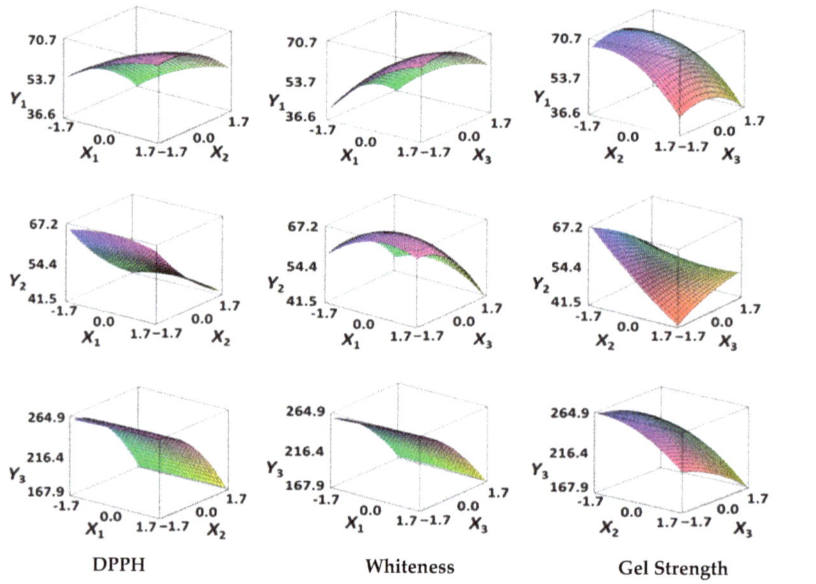

Figure 1. Three-dimensional response surface graph illustrating the correlation between independent variables (X1, X2, and X3) and dependent variables (Y1, Y2, and Y3). X1: Amount of antioxidant added (%); X2: Temperature conditions during the high-temperature and high-pressure treatment process (°C); X3: Time conditions in the high-temperature and high-pressure treatment process (min); Y1: DPPH (%); Y2: Whiteness; Y3: Gel strength (g × cm).

The lack of fit test, used to assess the suitability of the response model equation for the dependent variable, yielded p values of 0.095, 0.083, and 0.073 for DPPH (Y_1), whiteness (Y_2), and gel strength (Y_3), respectively, all of which are close to 1, indicating that the design model was appropriate [40] (Table 4).

Table 4. Estimated coefficients and p values of the second-order regression equation for optimization of manufacturing conditions for room-temperature fried fish cake prototype.

Variable	Y_1		Y_2		Y_3	
	Coefficient	p Value	Coefficient	p Value	Coefficient	p Value
Constant	63.197	0.000	53.357	0.000	236.807	0.000
X_1	4.311	0.000	0.061	0.847	0.263	0.708
X_2	−3.975	0.000	−3.809	0.000	−15.945	0.000
X_3	−1.414	0.025	−1.108	0.008	−12.640	0.000
X_1X_1	−0.954	0.124	−0.376	0.301	−0.097	0.900
X_2X_2	−1.626	0.021	0.473	0.203	−4.233	0.001
X_3X_3	−1.290	0.050	−0.446	0.227	−2.254	0.019
X_1X_2	0.175	0.795	0.063	0.880	−0.113	0.902
X_1X_3	−0.375	0.581	−0.213	0.611	−0.287	0.753
X_2X_3	0.025	0.970	1.938	0.002	−1.038	0.277

X_1: Amount of antioxidant added (%); X_2: Temperature conditions during the high-temperature and high-pressure treatment process (°C); X_3: Time conditions in the high-temperature and high-pressure treatment process (min); Y_1: DPPH (%); Y_2: Whiteness; Y_3: Gel strength (g × cm); X_1X_1, X_2X_2, X_3X_3, X_1X_2, X_1X_3, and X_2X_3: Coefficients of the quadratic regression equation.

Conditions involving high-temperature and high-pressure treatment of fried fish cake prototypes enhance safety through complete sterilization when treated at higher temperatures and for longer periods. However, excessive treatment may cause the deterioration of the quality of the fish cake. Additionally, the antioxidant effect may decrease due to heat and pressure. Considering these aspects, an appropriate time and temperature range is needed for manufacturing a high-quality and microbiologically safe fried fish cake prototype. Consequently, the desired values of the dependent variables [DPPH (Y_1), whiteness (Y_2), and gel strength (Y_3)] for determining optimal high-temperature and high-pressure treatment conditions were set to a maximum value of 70% for DPPH, 53.0 for whiteness, and 230.0 g × cm for gel strength based on the experimental results (Table 5).

Table 5. Optimal conditions predicted for the production of room-temperature fried fish cake prototype.

Independent Variable	Value	X_1		X_2		X_3	
Y_1	TV	70.0	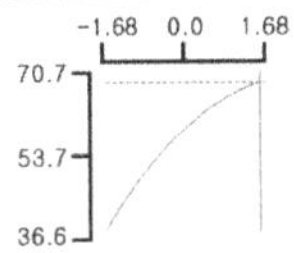	70.0	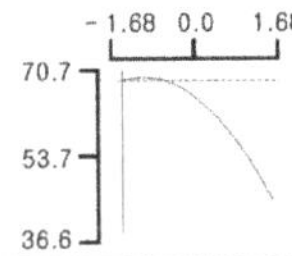	70.0	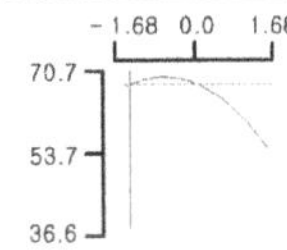
	CV	1.68		−1.63		−1.62	
	AC	0.18		106.9		16.9	
Y_2	TV	53.0	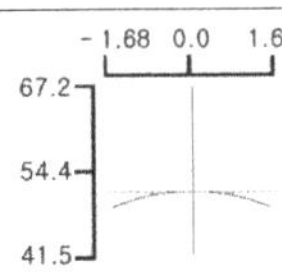	53.0	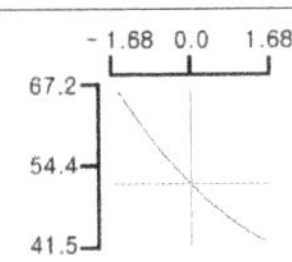	53.0	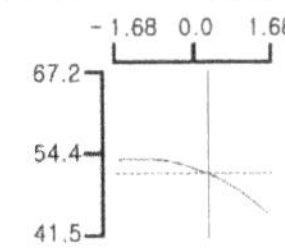
	CV	0.00		0.00		0.29	
	AC	0.10		115.0		30.5	
Y_3	TV	230.0	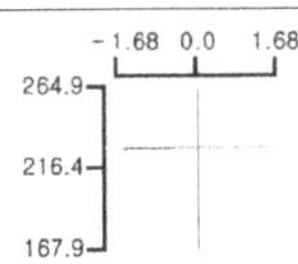	230.0	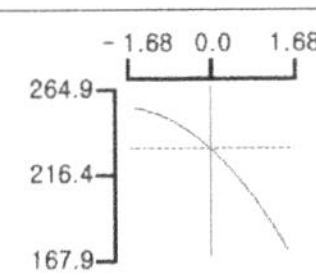	230.0	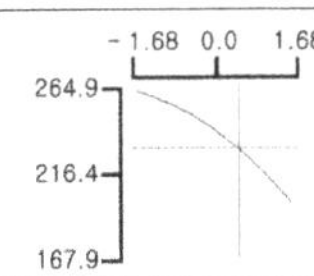
	CV	0.00		0.00		0.49	
	AC	0.10		115.0		27.5	
Optimization conditions for all dependent variables	CV	1.08		−0.49		0.75	
	AC	0.15%		112.6 °C		28.8 min	

X_1: Amount of antioxidant added (%); X_2: Temperature conditions during the high-temperature and high-pressure treatment process (°C); X_3: Time conditions in the high-temperature and high-pressure treatment process (min); Y_1: DPPH (%); Y_2: Whiteness; Y_3: Gel strength (g × cm); TV: Target value; CV: Coded value; AC: Actual experimental value.

The optimal amount of antioxidant added and the sign values for high-temperature and high-pressure treatment temperature and time that satisfied all dependent variables, considering the target values of independent variables for the fried fish cake prototype, were 1.08, −0.49, and 0.75, respectively, with actual values of 0.15%, 112.6 °C, and 28.8 min (Table 6).

Table 6. Predicted value and experimental value of various dependent variables derived under optimal conditions.

Dependent Variable	Predicted Value	Experimental Value
Y_1 (DPPH, %)	66.1 [a]	65.2 ± 1.7 [a]
Y_2 (Whiteness)	52.9 [a]	52.8 ± 0.9 [a]
Y_3 (Gel strength, g × cm)	233.2 [a]	235.1 ± 5.9 [a]

Values are means ± S.D. of three replications (n = 3). Y_1: DPPH (%); Y_2: Whiteness; Y_3: Gel strength (g × cm). Different superscript letters (a) indicate significant differences among means by Duncan's test ($p < 0.05$).

2.3. Nutritional Properties of Fried Fish Cake Prototype

2.3.1. Proximate Composition, pH, and Energy

The developed fried fish cake prototype had the highest moisture content at 72.4 g/100 g, which was similar to that of control group 3 but significantly different from control groups 1 and 2 (Table 7). In terms of protein, control groups 1 and 2 had a slightly higher content than the developed fried fish cake prototype, likely because the prototype had a high moisture content and a relatively low protein content (10.3 g/100 g). There was no significant difference in fat content among all products. Control group 2 had the highest ash content, while control groups 1 and 2 had the highest carbohydrate content. There was no significant difference in pH among all samples. Control groups 1 and 2, with a high carbohydrate content, showed high energy values.

Table 7. Proximate composition, pH, and energy of the controls and the developed fried fish cake prototype.

Sample	Proximate Components (g/100 g)					pH	Energy (kcal/100 g)
	Moisture	Protein	Fat	Ash	Carbohydrate		
Control 1	57.3	12.5	2.3	3.4	24.5	6.6	168.7
Control 2	56.8	13.0	2.2	4.1	23.9	6.2	167.4
Control 3	71.9	10.5	1.3	1.9	14.4	6.9	111.3
Fried fish cake prototype	72.4	10.3	1.4	1.9	14.0	6.9	109.8

Ash = 100 − (Moisture + Protein + Fat + Carbohydrates); Energy = [(Protein × 4.0) + (Fat × 9.0) + (Carbohydrates × 4.0)].

2.3.2. Total Amino Acid

The total amino acid content of the fried fish cake prototype was 10.17 g/100 g, with essential amino acids constituting 5.07 g/100 g (49.9%). The major amino acids (more than 8% of the total amino acids) in the fried fish cake prototype were aspartic acid (10.7%), glutamic acid (16.3%), leucine (8.9%), and lysine (9.4%) (Table 8). The variation in amino acid content between the fried fish cake prototype and the three control groups was attributed to variations in protein content resulting from the mixing conditions, while the proportions of each amino acid remained largely consistent.

Table 8. Total amino acid content of control and the developed fried fish cake prototype.

Amino Acids	RDR (g/70 kg Body Weight)	Control 1	Control 2	Control 3	FFCP
Histidine	1.6	0.31 (2.5)	0.26 (2.3)	0.20 (2.1)	0.23 (2.3)
Isoleucine	3	0.56 (4.5)	0.50 (4.4)	0.44 (4.7)	0.52 (5.1)
Leucine	6.1	1.03 (8.3)	0.94 (8.3)	0.81 (8.6)	0.90 (8.9)
Lysine	4.8	1.04 (8.4)	0.94 (8.3)	0.95 (10.1)	0.96 (9.4)
SAA	2.3	0.53 (4.3)	0.52 (4.6)	0.42 (4.5)	0.50 (4.9)
AAA	4.1	0.86 (7.0)	0.80 (7.0)	0.70 (7.4)	0.80 (7.8)
Threonine	2.5	0.56 (4.5)	0.52 (4.6)	0.45 (4.8)	0.48 (4.7)
Valine	4	0.6 (4.8)	0.56 (4.9)	0.46 (4.9)	0.54 (5.3)
∑EAA	28.4	5.49 (44.3)	5.04 (44.4)	4.43 (47.1)	4.93 (48.4)
Aspartic acid		1.14 (9.2)	1.05 (9.2)	0.97 (10.3)	1.09 (10.7)
Serine		0.53 (4.3)	0.49 (4.3)	0.42 (4.4)	0.42 (4.1)
Glutamic acid		2.82 (22.8)	2.59 (22.8)	1.78 (18.8)	1.66 (16.3)
Proline		0.60 (4.8)	0.55 (4.8)	0.29 (3.1)	0.34 (3.3)
Glycine		0.48 (3.9)	0.45 (4.0)	0.42 (4.4)	0.46 (4.5)
Alanine		0.67 (5.3)	0.61 (5.4)	0.54 (5.7)	0.61 (6.0)
Arginine		0.70 (5.6)	0.60 (5.3)	0.60 (6.4)	0.66 (6.5)
∑NAA		6.94 (55.9)	6.34 (55.8)	5.02 (53.1)	5.24 (51.4)

RDR: Recommended Daily Requirement according to Protein and Amino Acid Requirements in Human Nutrition [41]. FFCP: Fried fish cake prototype; SAA: Sulfur amino acids (cysteine, methionine). AAA: Aromatic amino acids (tyrosine, phenylalanine). ∑EAA: Sum of essential amino acids. ∑NAA: Sum of nonessential amino acids.

2.3.3. Minerals

The mineral composition of surimi and fish cakes is an essential criterion for surimi products and has a direct impact on consumer preferences [42]. The calcium content of the developed fried fish cake prototype was 35.3 mg/100 g, phosphorus was 116.0 mg/100 g, potassium was 66.4 mg/100 g, iron was 0.05 mg/100 g, and magnesium was 14.1 mg/100 g (Table 9). The mineral content of the control groups and the original product was measured, revealing differences in the mineral content of the fish cake samples depending on the type of surimi used as the main ingredient. This trend was consistent with the findings of [43], who observed that golden-threadfin bream surimi had a higher content of all minerals compared to pollack surimi, presumably due to the intrinsic mineral content differences in the surimi itself [43].

Table 9. Mineral content of controls and the developed fried fish cake prototype.

Mineral (mg/100 g)	Samples			
	Control 1	Control 2	Control 3	FFCP
Ca	463.5	470.5	33.4	35.3
P	192.5	196.9	106.9	116.0
K	139.8	139.4	65.6	66.4
Fe	0.75	0.73	0.10	0.05
Mg	21.6	21.6	13.5	14.1

Ca: Calcium; P: Phosphorus; K: Potassium; Fe: Iron; Mg: Magnesium; FFCP: Fried Fish Cake Prototype.

2.3.4. Fatty Acids

The total fatty acid content of the fried fish cake prototype was 1.03 g/100 g, which constituted 73.6% of the crude fat content (Table 10). In the samples analyzed, linoleic acid was found in significantly lower amounts in the fried fish cake prototype compared to the control groups. Margaric acid, arachidic acid, behenic acid, γ-linolenic acid, and erucic acid were present in small quantities in the control samples but were absent in the fried fish cake prototype. Conversely, elaidic acid was detected only in the fried fish cake prototype in trace amounts, not in the controls. These differences in fatty acid composition between the prototype and the controls are likely due to variations in manufacturing composition and processing conditions. Typically, during fish cake production, the tender fish meat is washed to remove contaminants such as lipids, pigments, and blood, resulting in low lipid content. The crude fat and fatty acids found in both the fried fish cake prototype and the control groups are attributed to the soybean oil used in the frying process.

Table 10. Fatty acid content in the controls and the developed fried fish cake prototype.

Fatty Acids (g/100 g)		Samples			
		Control 1	Control 2	Control 3	FFCP
C4:0	Butyric acid	-	-	-	-
C6:0	Caproic acid	-	-	-	-
C8:0	Caprylic acid	-	-	-	-
C10:0	Capric acid	-	-	-	-
C12:0	Lauric acid	-	-	-	-
C13:0	Tridecanoic acid	-	-	-	-
C14:0	Myristic acid	0.02	0.02	0.01	0.01
C15:0	Pentadecanoic acid	-	0.01	-	-
C15:1	Pentadecenoic acid	-	-	-	-
C16:0	Palmitic acid	0.33	0.35	0.17	0.15
C17:0	Margaric acid	0.01	0.01	-	-
C18:0	Stearic acid	0.10	0.11	0.06	0.05

Table 10. *Cont.*

Fatty Acids (g/100 g)		Samples			
		Control 1	Control 2	Control 3	FFCP
C20:0	Arachidic acid	0.01	0.01	-	-
C21:0	Heneicosanoic acid	-	-	-	-
C22:0	Behenic acid	0.01	0.01	0.01	-
C23:0	Tricosanoic acid	-	-	-	-
C24:0	Lignoceric acid	-	-	-	-
C14:1n-5	Myristoleic acid	-	-	-	-
C16:1n-7	Palmitoleic acid	0.02	0.02	-	0.01
C17:1	Margaroleic acid	-	-	-	-
C18:1t	Elaidic acid	-	-	-	0.01
C18:1c	Oleic acid	0.45	0.40	0.28	0.26
C20:1n-9	cis-11-Eicosenoic acid	0.01	0.01	0.01	0.01
C18:2t	Linolelaidic acid	-	-	-	-
C18:2c	Linoleic acid	0.95	0.84	0.53	0.36
C18:3γ	γ-Linolenic acid	0.01	0.01	0.01	-
C18:3n-3	α-Linolenic acid	0.08	0.07	0.05	0.03
C20:2n-6	cis-11,14-Eicosadienoic acid	-	-	-	-
C20:3n-6	dihomo γ-Linolenic acid	-	-	-	-
C20:3n-3	cis-11,14,17-Eicosatrienoic acid	-	-	-	-
C20:4n-6	Arachidonic acid	0.02	0.03	-	0.01
C20:5n-3	Eicosapentaenoic acid	0.03	0.04	0.04	0.04
C22:1n-9	Erucic acid	0.01	0.01	0.01	-
C22:2n-6	Docosadienoic acid	-	-	-	-
C22:6n-3	Docosahexaenoic acid	0.12	0.16	0.08	0.09
C24:1n-9	Nervonic acid	-	-	-	-
	Total fatty acids	2.18	2.11	1.26	1.03

-: Not Detected; FFCP: Fried fish cake prototype.

2.3.5. Odor Intensity and Volatile Basic Nitrogen (VBN)

The odor intensity of the developed fried fish cake prototype was 544.1 ± 14.35, with control group 2 having the highest intensity at 969.30 ± 7.56 and control group 3 the lowest at 535.40 ± 6.26. The VBN content of the developed prototype was 1.1 mg/100 g, with control group 2 being the highest at 3.1 mg/100 g, and control group 3 the lowest at 1.0 mg/100 g, following the same trend as odor intensity (Table 11). It was determined that the fishy odor of threadtail sea bream surimi used in control groups 1 and 2 was more pronounced, affecting both the odor intensity and VBN content. The increased odor intensity in control group 2 compared to control group 1 and the developed prototype suggests that heat treatment increased substances contributing to odor intensity. Despite heat treatment, the fishy smell was masked by the herbal oil addition, resulting in high sensory evaluation scores for flavor in both control group 3 and the developed prototype, which was consistent with overall preference trends.

Table 11. Odor intensity, volatile basic nitrogen, acid value, and Hunter color value of control and the developed fried fish cake product.

Experimental Items	Control 1	Control 2	Control 3	FFCP
Odor intensity (level)	864.8 ± 8.62 ^{b}	969.3 ± 7.56 ^{c}	535.4 ± 6.26 ^{a}	544.11 ± 14.35 ^{a}
VBN (mg/100 g)	2.8	3.1	1.0	1.1
Acid value (mg/100 g)	14.7	21.3	0.2	0.0
L value	47.03 ± 3.92 ^{a}	58.21 ± 2.24 ^{b}	77.50 ± 0.65 ^{c}	76.47 ± 0.53 ^{c}
a value	0.35 ± 0.06 ^{b}	10.96 ± 0.87 ^{c}	−2.76 ± 0.05 ^{a}	−2.29 ± 0.02 ^{a}
b value	15.32 ± 0.89 ^{b}	30.48 ± 0.85 ^{c}	6.81 ± 0.07 ^{a}	7.55 ± 0.09 ^{a}
Whiteness	21.08 ± 3.15 ^{b}	−33.24 ± 3.40 ^{a}	57.06 ± 0.46 ^{c}	53.83 ± 0.31 ^{c}

Table 11. *Cont.*

Experimental Items	Control 1	Control 2	Control 3	FFCP
ΔE value	52.56 ± 1.52 [a]	56.65 ± 2.25 [a,b]	59.00 ± 3.78 [b]	58.23 ± 1.66 [b]
Gel strength (g × cm)	231.80 ± 12.79 [a]	256.93 ± 3.74 [b]	244.40 ± 6.92 [a,b]	279.26 ± 15.42 [c]

Values are means ± S.D. of three replications (n = 3). L: Lightness; a: Redness; b: Yellowness; ΔE: Color difference; VBN: Volatile basic nitrogen; FFCP: Fried fish cake prototype. Different superscript letters (a–c) in each row indicate significant differences among means by Duncan's test ($p < 0.05$).

2.3.6. Acid Value

The acid value was not detected in the developed fried fish cake prototype at 0.01 mg/g (Table 11). This was attributed to the very low crude fat content and the prevention of oxidation through vacuum packing and subsequent heating. The Korean Standards (KS standards) for acid value in fried fish cake is 2.5 mg/g [44], and the developed prototype meets this standard.

2.3.7. Hunter Color Values

In the Hunter color values of the developed fried fish cake prototype, the L value was the highest at 76.47, with a and b values of −2.29 and 7.55, respectively, and whiteness at 53.83 (Table 11). This result was due to the absence of a reduction reaction of the added sugar, trehalose. The color preference of the developed prototype was the highest at 6.8 points, similar to control group 3 at 6.7 points, aligning with the overall preference trend. Generally, high whiteness is considered a marker of superior product quality [45]. Therefore, the addition of trehalose and optimized heating processes were used to minimize yellowness and enhance whiteness based on sensory evaluation results.

Control sample 2 showed a high b value, indicating increased yellowness, likely due to the Maillard reaction occurring during manufacturing, which involves a roasting stage. The ΔE value for the newly developed fried fish cake prototype indicated only slight color variation, attributed to the lower b values (yellowness) resulting from the addition of trehalose. Typically, fish cake products with higher whiteness are perceived as superior [46]. The increased whiteness in the fried fish cake prototype can be attributed to the incorporation of trehalose and optimized processing conditions. This prototype exhibited approximately 40% greater whiteness compared to control sample 1, however there was no significant difference when compared to control 3. Control sample 3 was produced under similar conditions but did not undergo sterilization. As a result, even though control 3 showed high whiteness, it was not significantly different from that of the developed fried fish cake prototype.

2.3.8. Gel Strength

The fried fish cake prototype demonstrated the highest gel strength, measuring 279.3 ± 15.42 g × cm (Table 11). Gel strength serves as a crucial quality indicator for fish cakes, where a higher breaking force suggests a firmer texture, and a greater penetration depth indicates higher elasticity. Therefore, a fish cake with significant penetration depth is considered to be of high quality.

Gel strength is a vital textural attribute that subsequently affects the quality and pricing of fish cake products [28,29,47]. It is calculated by multiplying the indentation strength (g) by the depth (cm), where indentation represents the applied force, and depth refers to the distance until this force peaks [48]. Generally, greater indentation strength indicates firmer physical properties, while a higher depth suggests that the force decreases gradually, reflecting high elasticity. The average indentation strength and depth values for the newly developed fried fish cake prototype and control groups 1, 2, and 3 were 331.14, 458.80, 305.50, and 362.67 g, and 0.70, 0.56, 0.80, and 0.77 cm, respectively. The developed fried fish cake prototype exhibited the highest gel strength among the samples, characterized by its highly elastic fish paste form and substantial depth.

2.4. Gel Electrophoresis

The SDS-PAGE analysis results of the developed fried fish cake prototype and control groups showed the myosin light chain of the prototype (D) and control group 3 (C) was most clearly identified in the 15–20 kDa range, followed by control group 2 (B) and control group 1 (A) (Figure 2). In the 20–25 kDa range, α-tropomyosin of the prototype (D) was identified. In the 25–37 kDa range, β-tropomyosin was most clearly identified in the prototype (D) and control 3 (C), followed by control 1 (A) and control 2 (B). In the 50–75 kDa range, G-actin was observed only in the prototype (D) and control 3 (C). In the 75–100 kDa range, heavy meromyosin was observed in control 3 (C). Among the peptide bands, the largest number appeared in the 37–50 kDa range, with various molecular weights confirmed overall in control 3 (C).

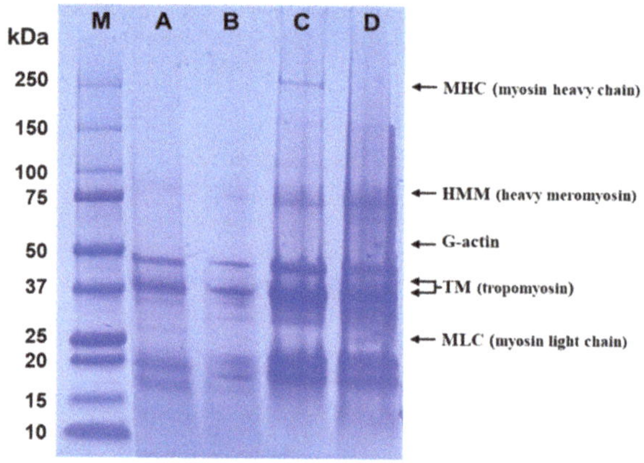

Figure 2. Protein distribution pattern in the developed fried fish cake prototype and the controls observed by SDS-PAGE. M: Molecular weight marker; A: Control 1; B: Control 2; C: Control 3; D: Developed fried fish cake prototype.

2.5. Microstructural Properties

The results of comparing the cross-sections of the developed fried fish cake prototype and control groups using scanning electron microscopy are shown in Figure 3. Control groups 1 and 2 exhibited very small pores and a clumped structure. Smaller pores indicate a weaker water retention capacity, affecting product elasticity. In contrast, control group 3 and the developed prototype had larger pores and a dense network structure, suggesting a better water retention capacity. This structural difference is presumed to be due to the presence or absence of the TGase enzyme, which maintains the protein network structure, and differences between threadtail sea bream meat (control groups 1 and 2) and pollack meat (control group 3 and the prototype).

The physical properties of the fish cake showed little difference in gel strength based on meat type and high-temperature and high-pressure treatment. However, microstructural observation confirmed that the regular network structure became less uniform, irregular, and clumped due to high-temperature and high-pressure treatment. Gel strength is the product of breaking force and penetration distance. In this experiment, no significant difference in gel strength was observed between the controls and the prototype. However, sensory evaluation showed the prototype with a uniform network structure had a superior texture and overall preference. This suggests that when subjected to high-temperature and high-pressure treatment, the added food additives, TGase and trehalose, prevented uneven protein network structure deterioration, ultimately increasing prototype preference.

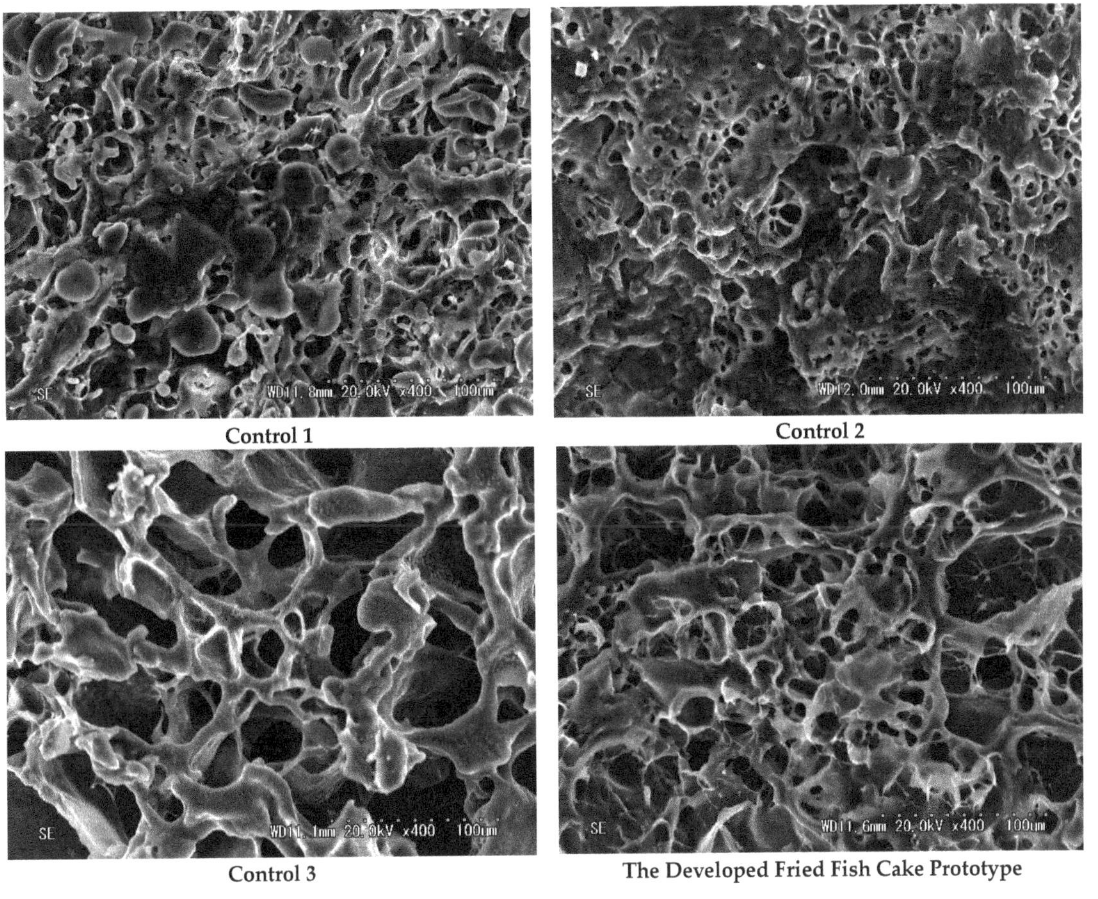

Figure 3. Scanning electron microscope results of controls and the developed fried fish cake prototype (Photo magnification: ×400).

2.6. Sensory Evaluation

The sensory evaluation results of the fried fish cake sample are presented in Table 12. The taste was rated at 6.8 ± 0.22 points, color at 6.6 ± 0.13 points, aroma at 6.3 ± 0.16 points, texture at 6.7 ± 0.11 points, and overall preference at 6.7 ± 0.15 points—all higher than control groups 1 and 2, and similar to control group 3.

Table 12. Sensory evaluation of control and the developed fried fish cake prototype using a 7-point scale method.

Sample	Sensory Evaluation (Score)				
	Taste	Color	Odor	Texture	Overall Acceptance
Control 1	5.0 ± 0.18 ^{b}	4.6 ± 0.20 ^{b}	4.5 ± 0.23 ^{b}	5.3 ± 0.17 ^{b}	5.2 ± 0.14 ^{b}
Control 2	4.4 ± 0.21 ^{a}	4.0 ± 0.24 ^{a}	3.5 ± 0.25 ^{a}	4.5 ± 0.10 ^{a}	4.8 ± 0.11 ^{a}
Control 3	6.8 ± 0.15 ^{c}	6.5 ± 0.15 ^{c}	6.2 ± 0.15 ^{c}	6.8 ± 0.12 ^{c}	6.7 ± 0.12 ^{c}
FFCP	6.8 ± 0.22 ^{c}	6.6 ± 0.13 ^{c}	6.3 ± 0.16 ^{c}	6.7 ± 0.11 ^{c}	6.7 ± 0.15 ^{c}

Values are means ± S.D. of three replications (n = 3). FFCP: Fried fish cake prototype. Different superscript letters (a–c) in each column indicate significant differences among means by Duncan's test ($p < 0.05$).

The newly developed fried fish paste cake prototype serves as a basic fish cake product, as it does not contain any additional auxiliary ingredients with strong flavor profiles such as vegetables or cheese. This allows for the customization of flavor by adding preferred auxiliary ingredients, even those with intense taste characteristics. The odor rating for this prototype was 6.3 ± 0.16, which was higher than that of grilled mackerel, grilled Spanish mackerel, grilled flounder, canned sardines, vegetable fish paste cake, and fish paste cake bars. In contrast, the control samples had a fishy smell, likely due to differences in frying, sterilization, and storage processes. Consequently, it was determined that the developed fried fish cake prototype possesses the most suitable odor score in comparison to the control samples. The prototype's appearance (color) was almost identical to the white color in the Munsell Table. Trehalose played a significant role in enhancing whiteness, indicating its usefulness in reducing color changes. The texture (elasticity) score of the prototype was 6.7 ± 0.11 points, surpassing that of acorn jelly, vegetable fish paste cake, konjac, and fish paste cake bars. This favorable result is due to the inclusion of TGase and the optimization of processing conditions involving high-temperature and high-pressure.

2.7. Shelf-Life Prediction

During 150 days of storage, the total bacterial counts (TBC), *Escherichia coli*, and coliform group, which are important indicators of food hygiene and microbiological characteristics according to storage temperature (25 °C, 35 °C, 45 °C), were not detected, indicating no bacterial presence at any storage temperature or period. *E. coli* and coliform bacteria were negative regardless of storage temperature and period.

The acid value and sensory changes, indicating fat oxidation degree according to storage temperature, are shown in Table 13. During storage, the acid value of the prototype ranged from 1.2 to 6.6 mg KOH/g at 25 °C, 1.2 to 10.2 mg KOH/g at 35 °C, and 1.2 to 12.6 mg KOH/g at 45 °C, all increasing with storage duration. Herbs are known to inhibit fat oxidation during food processing and storage [49]. The overall preference of sensory evaluation was measured using a 7-point scale, ranging from 6.1 to 7.0 points at 25 °C, 5.5 to 7.0 points at 35 °C, and 5.2 to 7.0 points at 45 °C, all decreasing over the storage period.

Table 13. Acid value (mg KOH/g) and sensory evaluation of the developed fried fish cake prototype stored at room-temperature for shelf-life prediction.

Storage Temperature	Day	Acid Value and Sensory Evaluation					
		Acid Value	Color	Odor	Taste	Texture	Overall Acceptance
25 °C	0	1.2 ± 0.1 ^{a}	7.0 ± 0.5 ^{e,f}	7.0 ± 0.5 ^{b,c}	7.0 ± 0.5 ^{c,d}	7.0 ± 0.5 ^{b,c}	7.0 ± 0.4 ^{c,d}
	30	2.7 ± 0.0 ^{b}	6.5 ± 0.0 ^{e}	6.8 ± 0.5 ^{b,c}	6.5 ± 0.0 ^{c}	6.5 ± 0.0 ^{b}	6.6 ± 0.2 ^{b,c}
	60	5.0 ± 0.0 ^{c}	6.3 ± 0.0 ^{d}	6.8 ± 0.5 ^{b,c}	6.5 ± 0.0 ^{c}	6.5 ± 0.0 ^{b}	6.6 ± 0.1 ^{b}
	90	5.4 ± 0.2 ^{d}	6.2 ± 0.0 ^{c}	6.2 ± 0.5 ^{a,b}	6.5 ± 0.0 ^{c}	6.3 ± 0.4 ^{a,b}	6.4 ± 0.2 ^{b}
	120	6.2 ± 0.2 ^{e}	5.7 ± 0.4 ^{a,b}	5.8 ± 0.4 ^{a}	6.2 ± 0.0 ^{b}	6.3 ± 0.4 ^{a,b}	6.2 ± 0.4 ^{a,b}
	150	6.2 ± 0.3 ^{e,f}	5.6 ± 0.2 ^{a}	5.8 ± 0.2 ^{a}	6.0 ± 0.2 ^{a}	6.1 ± 0.3 ^{a}	6.1 ± 0.3 ^{a}
35 °C	0	1.2 ± 0.1 ^{a}	7.0 ± 0.5 ^{c,d}	7.0 ± 0.5 ^{c,d}	7.0 ± 0.5 ^{c,d}	7.0 ± 0.5 ^{b,c}	7.0 ± 0.4 ^{b,c}
	30	1.9 ± 0.1 ^{b}	6.5 ± 0.0 ^{b,c}	6.6 ± 0.4 ^{c,d}	6.8 ± 0.5 ^{c,d}	6.8 ± 0.5 ^{b,c}	6.7 ± 0.3 ^{b,c}
	60	4.9 ± 0.1 ^{c}	6.3 ± 0.4 ^{b,c}	6.0 ± 0.5 ^{b,c}	6.5 ± 0.0 ^{c}	6.5 ± 0.0 ^{b}	6.3 ± 0.1 ^{b}
	90	8.4 ± 0.3 ^{d}	6.2 ± 0.5 ^{b,c}	5.7 ± 0.0 ^{b}	6.0 ± 0.5 ^{b,c}	6.2 ± 0.5 ^{a,b}	6.0 ± 0.1 ^{a,b}
	120	9.4 ± 0.3 ^{e}	5.7 ± 0.0 ^{b}	5.0 ± 0.4 ^{a}	5.7 ± 0.0 ^{b}	5.7 ± 0.0 ^{a}	5.5 ± 0.1 ^{a}
	150	10.7 ± 0.2 ^{f}	5.4 ± 0.1 ^{a}	4.9 ± 0.2 ^{a}	5.5 ± 0.1 ^{a}	5.6 ± 0.1 ^{a}	5.5 ± 0.2 ^{a}
45 °C	0	1.2 ± 0.1 ^{a}	7.0 ± 0.5 ^{c,d}	7.0 ± 0.5 ^{c,d}	7.0 ± 0.5 ^{c,d}	7.0 ± 0.5 ^{c,d}	7.0 ± 0.4 ^{c,d}
	30	6.0 ± 0.1 ^{b}	6.5 ± 0.0 ^{b,c}	6.5 ± 0.0 ^{c}	6.5 ± 0.0 ^{c}	6.6 ± 0.4 ^{c}	6.6 ± 0.1 ^{c}
	60	6.7 ± 0.1 ^{c}	6.3 ± 0.4 ^{b,c}	5.8 ± 0.4 ^{a,b}	5.8 ± 0.4 ^{b,c}	6.0 ± 0.5 ^{b,c}	6.0 ± 0.1 ^{b,c}
	90	9.6 ± 0.4 ^{d}	6.2 ± 0.5 ^{b,c}	5.2 ± 0.5 ^{a}	5.5 ± 0.4 ^{a,b}	5.7 ± 0.0 ^{b}	5.7 ± 0.2 ^{b}
	120	11.4 ± 0.4 ^{e}	5.7 ± 0.0 ^{b}	5.0 ± 0.4 ^{a}	5.2 ± 0.5 ^{a}	5.5 ± 0.4 ^{a,b}	5.3 ± 0.1 ^{a}
	150	13.2 ± 0.2 ^{f}	5.3 ± 0.1 ^{a}	4.8 ± 0.2 ^{a}	4.9 ± 0.2 ^{a}	5.3 ± 0.3 ^{a}	5.2 ± 0.2 ^{a}

Values are mean ± S.D. Different letters (a–f) in each column at each temperature indicate significant differences among means by Duncan's test ($p < 0.05$).

The expected shelf-life of the developed fried fish cake prototype is shown in Table 13. Park et al. [50] described that a model with a high R^2 value is suitable for predicting shelf-life. In this study, the predicted shelf-life (23.93 months) was determined using the 0th model results, showing a higher R^2 than the 1st model at each storage temperature. However, multiple factors influence shelf-life conditions independent of how the product is handled during production, distribution, and storage [45]. Therefore, the final shelf-life was determined to be approximately 19 months after applying a safety factor of 0.8. This is approximately 11 times longer than the commercially available control group 1, which has a shelf-life of approximately 50 days (Table 14).

Table 14. Shelf-life estimation using accelerated testing (utilizing the Shelf-Life Prediction program provided by the Korean MFDS) [51].

Determined Parameter	Response Order	Temperature (°C)	R^2	Predicted Shelf-Life (Month)
Acid value	Zero order	25	0.9583	23.93
		35	0.9377	
		45	0.9782	
	First order	25	0.9040	9.70
		35	0.9032	
		45	0.8489	

MFDS: Korean Ministry of Food and Drug Safety; R^2: Coefficient of determination.

3. Conclusions

This research successfully developed a fried fish cake prototype with enhanced quality attributes and an extended shelf-life, suitable for room-temperature distribution. By integrating TGase, trehalose, and bay oil, and optimizing the heating process via the response surface methodology, the study effectively addressed the challenges associated with high-temperature and high-pressure treatments. TGase played a crucial role in enhancing gel strength through protein cross-linking, while trehalose improved color stability by reducing browning, thereby increasing the product's visual appeal. Bay oil was instrumental in enhancing the flavor profile by effectively masking undesirable odors, ensuring a pleasant sensory experience. The optimized processing conditions not only maximized antioxidant activity and maintained desirable whiteness and gel strength but also ensured compliance with nutritional standards, including a balanced composition of moisture, protein, essential amino acids, and minerals. The uniform microstructure observed in the prototype contributed to its superior texture and sensory evaluations. Furthermore, the shelf-life prediction indicated a storage duration of approximately 19 months, offering a competitive edge over existing products. This innovative approach not only facilitates room-temperature storage and distribution but also paves the way for broader applications of these techniques in surimi-based product development, potentially increasing profitability and market reach.

4. Materials and Methods

4.1. Materials

The development of the room-temperature fried fish cake prototype utilized the following materials: Alaskan pollock (*Gadus chalcogrammus*) surimi (AA grade; Westward Seafood Inc., Bellevue, WA, USA), transglutaminase (ACTIVA TGase-K, Ajinomoto, Tokyo, Japan), trehalose (Tongliao Meihua Biological Technology Co., Ltd., Langfang, China), D-xylose (Healtang Biotech Co., Ltd., Jinan, China), glucose (Q 1 Co., Ltd., Seoul, Republic of Korea), sucrose (Cheil Jedang Co., Ltd., Seoul, Republic of Korea), herb oils [bay tree (Laurel Noble Oil, Dotterline Co., Ltd., Hwaseong, Republic of Korea), aniseed, coriander, cumin, basil oil (Moellhausen Co., Vimercate, MB, Italy)], tocopherol (mixed tocopherol 70%; Namyung Co., Ltd., Seoul, Republic of Korea), ice (Handong Co., Ltd., Busan, Republic of Korea), refined salt (Solar Salt Co., Ltd., Gwangju, Republic of Korea), fish meat extract

(Jin-sung FM Co., Hwaseong, Republic of Korea), monosodium glutamate (Miwo, Daesang Co., Ltd., Seoul, Republic of Korea), glycine (Hebei Huayang Biological Technology Co., Ltd., Hengshui, China), egg white powder (Eurovo SRL Co., Ltd., Emilia-Romagna, Italy), phosphate complex (Polymix, Seodo BNI Co., Ltd., Hwaseong, Republic of Korea), tapioca starch (Moafood Co., Ltd., Gwangju, Republic of Korea), and potato starch (Dongafood Co., Ltd., Ulsan, Republic of Korea).

4.2. Manufacturing of Fried Fish Cake Prototype

The mixing ratio of main and auxiliary ingredients for the fried fish cake prototypes followed the guidelines outlined by Lee et al. [4], as detailed in Table 1. In brief, the dough was prepared by mixing the main and auxiliary ingredients, then steamed in a steamer machine (CHDC–500; Chungha Machinery Co., Ltd., Busan, Republic of Korea) at 47 ± 1 °C for 10 min. Subsequently, the steamed fish cake was fried in a fryer (WS-EF S20; Woosung Co., Ltd., Seoul, Republic of Korea) at 160 ± 3 °C for 6 min and cooled to room temperature for 10 min. The cooled fried fish cake was vacuum-packed in a retort pouch (PET12/AL9/NY15/CPP80) using a vacuum packer (TPS-V75; MAPPACK Co., Hwaseong-si, Republic of Korea), and subjected to high-temperature (106.6–123.4 °C) and pressure (1.3 kg/cm^2) treatment for 16.6–33.4 min using a retort sterilizer (PRS-06-I; Kyunghan Co., Ltd., Gyeongsan, Republic of Korea).

4.3. Optimizing Processing Conditions of Fried Fish Cakes Prototype

4.3.1. Optimization of TGase and Sugar Ratios to Improve Texture and Color

The impact of different sugars on the gel strength and chromaticity of fried fish cake prototypes was assessed. Sugars tested included sucrose, glucose, xylose, and trehalose, each at 0.75% concentration. TGase (ACTIVA TGase-K) was added at 0.37%. The prototypes underwent steaming at 55 ± 1 °C for 10 min, followed by oil frying at 160 ± 3 °C for 3 min. The semi-prototypes were cooled to room temperature, vacuum-packed in retort film, and autoclaved at 115 °C for 25 min. Their physical properties were evaluated using a texture analyzer. Yellowness and whiteness measurements were conducted using color difference meter (CM-700d; Konica Minolta Sensing lnc., Tokyo, Japan) to identify the most suitable sugars for the prototype.

4.3.2. Improving Flavor with Various Herbal Oils

To enhance the flavor of the fried fish cake prototypes, various herb oils (bay tree, aniseed, coriander, cumin, and basil oil) were incorporated at 0.1% during the dough mixing process. The intensity of odors in different prototypes was measured using color difference meter (CM-700d; Konica Minolta Sensing lnc., Tokyo, Japan), and sensory evaluations were conducted as described in Section 4.8.

4.4. Response Surface Regression Analysis

Optimal conditions for high-temperature and high-pressure processing of the prototypes were determined using dependent variables: DPPH radical scavenging activity (Y_1) for antioxidant activity, whiteness (Y_2) for color change, and gel strength (Y_3) for elasticity. A regression analysis was conducted using the MINITAB statistical software (version 14, MINITAB, State College, PA, USA). The relationships between dependent and independent variables were graphed using MAPLE software (version 12, Maple Soft, Waterloo, ON, Canada). The graph showing the relationship between the independent and dependent variables was expressed as a three-dimensional graph by inputting the values of the constant, first-order terms, second-order terms, and cross-terms, which are the outcomes derived from the regression equation based on the results of the aforementioned regression analysis. Optimal conditions were derived following the methods of Kim et al. [52] and Bezerra et al. [53], utilizing the response optimizer in MINITAB.

4.5. Preparation of Controls for Evaluation of Quality Characteristics

To compare the quality characteristics of the final developed fried fish cake prototypes, controls 1 and 2 were prepared using commercial products (Samjin Food Co., Ltd., Busan, Republic of Korea). Controls 1 and 2 included frozen red snapper (74.24%) with additional ingredients like wheat flour, refined salt, xylose, soybean oil, and others. Control 1 was produced without retorting, while control 2 underwent retorting under the same conditions as the developed prototype. Control 3 was prepared identically to the developed prototype but was non-sterilized.

4.6. Quality Analysis of Prepared Fried Fish Cake Prototype

4.6.1. Proximate Composition Analysis

The Proximate composition analysis followed the Food Codex method [51]. Carbohydrates were determined by the formula: [100 − (moisture + crude protein + crude fat + ash content)]. Energy was calculated using the FAO/WHO energy conversion factors: [(protein × 4.0) + (fat × 9.0) + (carbohydrate × 4.0)] [54].

4.6.2. pH

The pH value was measured by homogenizing approximately 5 g of the crushed fried fish cake with distilled water (9 times v/w) using a Homogenizer (SHG-15D; SciLab, Seoul, Republic of Korea). The resulting mixture was then centrifuged at 3000 rpm for 20 min using a centrifuge (Fleta 5; Hanil Scientific Inc., Gimpo, Republic of Korea). The pH was subsequently determined with a pH meter (Ohaus Starter 2100; Ohaus Co., Parsippany, NJ, USA).

4.6.3. Volatile Basic Nitrogen (VBN)

Volatile Basic Nitrogen (VBN) was measured following the method reported in Food Codex [51]. The supernatant obtained from a homogenized mixture was centrifuged at $1100\times g$ for 10 min and then filtered through a filter (TY5A-150; Advantec Toyo Kaisha LTD., Tokyo, Japan) and used for VBN measurement, which was calculated using Equation (1).

$$\text{VBN (mg\%)} = 0.14 \times \{(\text{sample titration} - \text{control titration}) \times \text{f}\}/\text{W} \times 100 \times 50 \quad (1)$$

where f stands for factor of 0.01 N NaOH and W indicates the amount of sample (g).

4.6.4. Volatile Component Intensity (VCI)

Volatile Component Intensity (VCI) was assessed using the method by Kang et al. [55]. In summary, approximately 5 g of fried fish cake sample was placed in a 50 mL conical tube (SPL Life Science Co., Ltd., Gyeonggi, Republic of Korea) and sealed with Parafilm (Heathrow Scientific, Lakeview Parkway, IL, USA) to prevent odor loss. VCI was monitored with an electronic nose (Odor concentration meter, XP-329R; New Cosmos Electric Co., Ltd., Osaka, Japan) set to batch mode, and the odor intensity was expressed in levels.

4.6.5. Measurement of Acid Value

The acid value was determined using the method outlined in the Food Codex [51].

4.6.6. Color Characteristics

The Hunter color values, including lightness (L), redness (a), yellowness (b), and color difference (ΔE) of the fish cake prototype, were recorded with a color difference meter (Konica Minolta Sensing Inc., Tokyo, Japan) following the method by Lee et al. [4]. For reference, the standard white plate had L, a, and b values of 20.69, −4.59, and −9.01, respectively. The color difference (ΔE) was estimated using the formula in Equation (2).

$$\Delta E = [(\Delta \text{L})^2 + (\Delta \text{a})^2 + (\Delta \text{b})^2]^{1/2} \quad (2)$$

The sample for whiteness measurement was cut into 1 cm × 1 cm (length and width) sections from the inner part of the fried fish cake prototype. After measuring L and b values, the whiteness value was calculated by L − 3b as described by Park [56].

4.6.7. Measurement of Gel Strength

Gel strength was assessed using a texture analyzer (CT3-1000; Brookfield, Middleboro, MA, USA) following the method described by Lee et al. [4]. Briefly, the gel strength of the fried fish cake samples, cut into dimensions of 10 × 10 × 4 mm, was measured using a circular plunger with a 14 mm diameter. The measurement conditions included a distance of 1.5 mm, a test speed of 0.5 mm/s, and a trigger load of 4.0 g.

4.7. Antioxidant Activity Analysis

The DPPH radical scavenging activity of the fried fish cake prototypes was calculated following the modified protocol of Teixeira et al. [57]. Five grams sample was mixed with 20 mL of methanol (5-fold dilution) and centrifuged at 3000× *g* for 20 min. The supernatant was filtered using a 0.45 μm pore-sized syringe filter (Hyundai Micro, Seoul, Republic of Korea). All processes were performed at 4 °C or below. The samples were mixed with a 0.2 mM DPPH-radical solution and left to react in the dark for 30 min. The absorbance was then measured at 517 nm using an absorbance meter (SPECTROstar Nano; BMG Labtech, Ortenberg, Hesse, Germany). The DPPH radical scavenging activity was calculated according to the formula of Kim et al. [31].

4.8. Sensory Evaluation Analysis

A descriptive evaluation was conducted to select the best herb oil suitable for the fried fish cake prototype using a 7-point rating method by five trained panel members (ages 20–50 years, 2 men and 3 women) affiliated with Samjin Food Co., Ltd. Busan, Republic of Korea. The descriptive analysis items were rated as follows: 7 points for very good, 5 points for average, and 1 point for very bad, evaluating appearance, smell, and overall preference.

The sensory evaluation of the final product prototype was performed using a 9-point scale by a panel of 21 people [8 males (maximum age: 50 years, minimum age: 22 years, average 35.8 ± 8.7) and 13 females (maximum age: 48 years, minimum age: 23 years, average 32.1 ± 7.2)] to evaluate taste, color, odor, texture, and overall acceptance of the final product. The study was conducted according to the guidelines of the Declaration of Helsinki and approved by the Institutional Review Board of Gyeongsang National University [protocol code: GIRB-E24-NY-0010].

4.9. Total Amino Acids, Mineral, and Fatty Acid Content

The total amino acids and mineral content of the fried fish cake prototype were measured using the method of the Food Code [51]. Sample oil for fatty acid analysis was extracted following the Bligh and Dyer protocol [58] with chloroform-methanol (2:1, *v*/*v*) used as the extraction solvent. Each analysis was performed with an amino acid auto-analyzer (Pharmacia Biotech Biochrom 30; Biochrom Ltd., Waterbeach Cambridge, UK), inductively coupled plasma optical emission spectrometry (ICP-OES), and gas chromatography (Shimadzu 14A; carrier gas, He; detector, FID) for identification and calculation.

4.10. Molecular Weight Analysis

The molecular weight of the proteins in the developed product prototype was determined using Sodium Dodecyl Sulfate Polyacrylamide Gel Electrophoresis (SDS–PAGE; 041BR97428; PowerPac Basic, Bio-Rad Laboratories, Inc., Hercules, CA, USA) following the method by Bashir et al. [59]. Briefly, each lane was loaded with 100 μg protein on precast 4–20% Mini-PROTEAN® TGX™ bis-Tris gels. The electrophoresis was performed in a Tris/glycine/SDS buffer within a Mini-PROTEAN tetra cell at a constant voltage of 100 V. The Precision Plus Protein Dual Color Marker (1610374; Bio-Rad) was used as a standard for molecular weight reference. The gel was stained with Coomassie Brilliant Blue R-250 solution for one hour, followed by destaining for 3 to 4 h. The protein bands were then visualized using the ViewOne Lablite system (EmbiTec, San Diego, CA, USA).

4.11. Scanning Electron Microscopy (SEM)

The microstructure of the developed fried fish cake prototype was observed according to the method of Park et al. [60]. Samples were thinly sectioned, freeze-dried, and coated with gold ions using ion sputtering technique (E-1010; Hitachi Instrument Inc., San Jose, CA, USA). The microstructure was then analyzed using a scanning electron microscope (JSM-6490LV; JEOL, Tokyo, Japan). Images were captured with an electron acceleration voltage of 15 kV, and the structure and morphology of the fish cake prototype were examined at low (×100 and ×200) and high (×400) magnifications. Image analysis was conducted using SigmaScan Pro version 5 (Systat Software Inc., San Jose, CA, USA).

4.12. Shelf-Life Analysis

The expiration date of the developed prototype was predicted using the Visual Shelf-Life Simulator for Foods (VSLSF), a food expiration date calculation system provided by the Food Codex [51]. To accurately predict the expiration date, an accelerated test was conducted at three storage temperatures (25, 35, and 45 °C). Quality indicators such as general bacterial count, *Escherichia coli*, acid value, and sensory test results were input into the VSLSF. The consumption period of the prototype was determined by multiplying the predicted value by a safety factor of 0.8.

4.13. Statistical Analysis

All experimental analyses were performed in triplicate. The results are expressed as means ± standard deviation (S.D.). Statistical analyses were conducted using ANOVA in SPSS Statistics version 18 (IBM Corp., Armonk, NY, USA). Post hoc comparisons were made using Duncan's multiple range test, with statistical significance set at $p < 0.05$.

Author Contributions: Conceptualization, K.M.I.B. and J.-S.C.; methodology, Y.Y.K., K.M.I.B., J.H.S. and J.-S.C.; software, K.M.I.B. and J.-S.C.; validation, K.M.I.B., J.H.S. and J.-S.C.; formal analysis, Y.Y.K., K.M.I.B., S.K. and Y.P.; investigation, Y.Y.K., K.M.I.B. and J.-S.C.; resources, J.-S.C.; data curation, Y.Y.K., K.M.I.B., S.K. and Y.P.; writing—original draft preparation, Y.Y.K., K.M.I.B. and J.-S.C.; writing—review and editing, K.M.I.B., J.H.S. and J.-S.C.; visualization, K.M.I.B., J.H.S. and J.-S.C.; supervision, J.-S.C.; project administration, J.-S.C.; funding acquisition, J.-S.C. All authors have read and agreed to the published version of the manuscript.

Funding: This research was supported by the Korea Institute of Marine Science & Technology Promotion (KIMST) and was funded by the Ministry of Oceans and Fisheries (20220272).

Institutional Review Board Statement: Not applicable.

Informed Consent Statement: Not applicable.

Data Availability Statement: The original contributions presented in the study are included in the article, further inquiries can be directed to the corresponding authors.

Conflicts of Interest: Author Ye Youl Kim, Seonyeong Kang and Yongjoon Park were employed by the company Samjin Food Co. Ltd. The authors declare that the research was conducted in the absence of any commercial or financial relationships that could be construed as a potential conflict of interest.

References

1. Kong, W.; Zhang, T.; Feng, D.; Xue, Y.; Wang, Y.; Li, Z.; Yang, W.; Xue, C. Effects of modified starches on the gel properties of Alaska Pollock surimi subjected to different temperature treatments. *Food Hydrocoll.* **2016**, *56*, 20–28. [CrossRef]
2. Shi, H.; Qin, R.; Wu, R.; Rong, J.; Jia, C.; Liu, R. Effect of cryoprotectants on the formation of advanced glycation end products and acrylamide in fried fish cakes. *Food Biosci.* **2021**, *44*, 101433. [CrossRef]
3. KMI—Korea Maritime Institute. Available online: https://kfishinfo.co.kr/kor/view.do?no=626 (accessed on 22 May 2024).
4. Lee, J.-H.; Kang, S.I.; Mansoor, S.; Lee, I.; Kim, D.Y.; Kim, Y.Y.; Park, Y.; Sohn, J.-H.; Bashir, K.M.I.; Choi, J.-S. Development and process optimization of a steamed fish paste cake prototype for room temperature distribution. *Processes* **2024**, *12*, 795. [CrossRef]

5. Kaewudom, P.; Benjakul, S.; Kijroongrojana, K. Properties of surimi gel as influenced by fish gelatin and microbial transglutaminase. *Food Biosci.* **2013**, *1*, 39–47. [CrossRef]
6. Luo, X.; Huang, K.; Lei, Y.; An, Y.; Xiong, S.; Hu, Y. Effects of freezing on quality attributes of surimi gels with various cross-linking degrees using MTGase: Quantitative analysis based on the ice crystals, network structure, and taste substances. *J. Food Eng.* **2024**, *371*, 111975. [CrossRef]
7. Liu, J.; Wang, X.; Ding, Y. Optimization of adding konjac glucomannan to improve gel properties of low-quality surimi. *Carbohydr. Polym.* **2013**, *92*, 484–489. [CrossRef]
8. Tan, Z.; Yang, X.; Wang, Z.; Chen, Z.; Pan, J.; Sun, Q.; Dong, X. Konjac glucomannan-assisted fabrication of stable emulsion-based oleogels constructed with pea protein isolate and its application in surimi gels. *Food Chem.* **2024**, *443*, 138538. [CrossRef]
9. Lan, H.; Chen, L.; Wang, Y.; Lu, M.; Chen, B.; Ai, C.; Teng, H. Effect of κ-carrageenan on saltiness perception and texture characteristic related to salt release in low-salt surimi. *Int. J. Biol. Macromol.* **2023**, *253*, 126852. [CrossRef]
10. Niu, Z.; Yu, J.; Xue, Y.; Xue, C.; Mao, X. Synergistic effects and interactions among components in the surimi/κ-carrageenan/soybean oil system. *Food Hydrocoll.* **2024**, *156*, 110250. [CrossRef]
11. Dong, X.; Pan, Y.; Zhao, W.; Huang, Y.; Qu, W.; Pan, J.; Qi, H.; Prakash, S. Impact of microbial transglutaminase on 3D printing quality of *Scomberomorus niphonius* surimi. *LWT-Food Sci. Technol.* **2020**, *124*, 109123. [CrossRef]
12. Liang, F.; Lin, L.; He, T.; Zhou, X.; Jiang, S.; Lu, J. Effect of transglutaminase on gel properties of surimi and precocious Chinese mitten crab (*Eriocheir sinensis*) meat. *Food Hydrocoll.* **2020**, *98*, 105261. [CrossRef]
13. Fang, Q.; Shi, L.F.; Ren, Z.Y.; Hao, G.X.; Chen, J. Effects of emulsified lard and TGase on gel properties of threadfin bream (*Nemipterus virgatus*) surimi. *LWT-Food Sci. Technol.* **2021**, *146*, 111513. [CrossRef]
14. Pietrowski, B.N.; Tahergorabi, R.; Jaczynski, J. Dynamic rheology and thermal transitions of surimi seafood enhanced with ω-3-rich oils. *Food Hydrocoll.* **2012**, *27*, 384–389. [CrossRef]
15. Hu, Y.; Shao, Y.; Wu, C.; Yuan, C.; Ishimura, G.; Liu, W.; Chen, S. γ-PGA and MTGase improve the formation of ε-(γ-glutamyl) lysine cross-links within hairtail (*Trichiurus haumela*) surimi protein. *Food Chem.* **2018**, *242*, 330–337. [CrossRef]
16. Lund, M.N.; Ray, C.A. Control of Maillard reactions in foods: Strategies and chemical mechanisms. *J. Agric. Food Chem.* **2017**, *65*, 4537–4552. [CrossRef]
17. Hofmann, T. Studies on the relationship between molecular weight and the color potency of fractions obtained by thermal treatment of glucose/amino acid and glucose/protein solutions by using ultracentrifugation and color dilution techniques. *J. Agric. Food Chem.* **1998**, *46*, 3891–3895. [CrossRef]
18. O'Brien, J. Stability of trehalose, sucrose and glucose to nonenzymatic browning in model systems. *J. Food Sci.* **1996**, *61*, 679–682. [CrossRef]
19. Chen, A.; Tapia, H.; Goddard, J.M.; Gibney, P.A. Trehalose and its applications in the food industry. *Compr. Rev. Food Sci. Food Saf.* **2022**, *21*, 5004–5037. [CrossRef]
20. Xu, S.; Li, P.; Han, F.; Zhou, H.; Zhou, K.; Wang, Y.; Cai, K.; Li, C.; Xu, B. Myofibrillar protein interacting with trehalose elevated the quality of frozen meat. *Foods* **2022**, *11*, 1041. [CrossRef]
21. Jain, N.K.; Roy, I. Effect of trehalose on protein structure. *Prot. Sci.* **2008**, *18*, 24–36. [CrossRef]
22. Huang, L.; Ruan, L.; Guo, Y.; Liang, X.; Shang, C. The effect of trehalose on the contents of starch and oil in microalga *Dunaliella parva*. *Algal Res.* **2024**, *81*, 103600. [CrossRef]
23. Babic, J.; Subaric, D.; Milicevic, B.; Ackar, D.; Kopjar, M.; Nedic, T.N. Influence of trehalose, glucose, fructose, and sucrose on gelatinisation and retrogradation of corn and tapioca starches. *Czech J. Food Sci.* **2009**, *27*, 151–157. [CrossRef]
24. Kim, H.Y.; Noh, K.S. Effect of trehalose on the shelf-life of backsulgies. *Korean J. Food Cookery Sci.* **2008**, *24*, 912–918.
25. Kang, H.J.; Kim, S.H.; Lim, J.K. Effect of trehalose on moisture and texture characteristics of instant baekseolgi prepared by microwave oven. *Korean J. Food Sci. Technol.* **2010**, *42*, 304–309.
26. Motoki, M.; Seguro, K. Transglutaminase and its use for food processing. *Trends Food Sci. Technol.* **1998**, *9*, 204–210. [CrossRef]
27. Lorenzen, P.C. Renneting properties of transglutaminase-treated milk. *Milchwissenschaft* **2000**, *55*, 433–437.
28. Bashir, K.M.I.; Kim, J.S.; An, J.H.; Sohn, J.H.; Choi, J.S. Natural food additives and preservatives for fish-paste products: A review of the past, present, and future states of research. *J. Food Qual.* **2017**, *2017*, 9675469. [CrossRef]
29. Benjakul, S.; Chantakun, K.; Karnjanapratum, S. Impact of retort process on characteristics and bioactivities of herbal soup based on hydrolyzed collagen from seabass skin. *J. Food Sci. Technol.* **2018**, *55*, 3779–3791. [CrossRef]
30. Jang, J.A.; Kim, H.A.; Choi, S.K. Quality characteristics of fish cake made with silver pomfret (*Pampus argenteus*) with added wasabi powder. *J. East Asian Soc. Diet. Life* **2010**, *20*, 103–112.
31. Kim, B.S.; Oh, B.J.; Lee, H.I. Comparison of the physicochemical properties and antioxidant activity of steamed fish paste containing seaweeds (sea mustard, laver, sea tangle) powder. *J. Korean Soc. Food Sci. Nutr.* **2020**, *49*, 1169–1174. [CrossRef]
32. Kuraishi, C.; Yamazaki, K.; Susa, Y. Transglutaminase: Its utilization in the food industry. *Food Rev. Int.* **2007**, *17*, 221–246. [CrossRef]
33. Santhi, D.; Kalaikannan, A.; Malairaj, P.; Arun Prabhu, S. Application of microbial transglutaminase in meat foods: A review. *Crit. Rev. Food Sci. Nutr.* **2017**, *57*, 2071–2076. [CrossRef] [PubMed]
34. Runglerdkriangkrai, J.; Banlue, K.; Raksakulthai, N. High temperature tolerant fish protein gel using transglutaminase and sodium ascorbate. *Kasetsart J. Nat. Sci.* **2006**, *40*, 84–90.

35. Monteiro, M.L.G.; Mársico, E.T.; Lázaro, C.A.; da Silva Canto, A.C.V.C.; da Costa Lima, B.R.C.; da Cruz, A.G.; Conte-Júnior, C.A. Effect of transglutaminase on quality characteristics of a value-added product tilapia wastes. *J. Food Sci. Technol.* **2015**, *52*, 2598–2609. [CrossRef]
36. Jain, N.K.; Roy, I. Trehalose and protein stability. *Curr. Protoc. Protein Sci.* **2010**, *59*, 4.9.1–4.9.12. [CrossRef]
37. Ryu, S.Y.; Roh, H.J.; Noh, B.S.; Kim, S.Y.; Oh, D.K.; Lee, W.J.; Yoon, J.R.; Kim, S.S. Effects of various sugars including tagatose and their molar concentrations on the Maillard browning reaction. *Korean J. Food Sci. Technol.* **2003**, *35*, 898–904.
38. Jiang, L.; Tian, Y.; Zhang, H.; Liu, S. Molecular-level insight into the effects of low moisture and trehalose on the thermostability of β-glucosidase. *Food Chem.* **2024**, *460*, 140607. [CrossRef]
39. Chiozzi, V.; Agriopoulou, S.; Varzakas, T. Advances, applications, and comparison of thermal (pasteurization, sterilization, and aseptic packaging) against non-thermal (ultrasounds, UV radiation, ozonation, high hydrostatic pressure) technologies in food processing. *Appl. Sci.* **2022**, *12*, 2202. [CrossRef]
40. Zhou, P.; Regenstein, J.M. Optimization of extraction conditions for pollock skin gelatin. *J. Food Sci.* **2004**, *69*, 393–398.
41. WHO. *Protein and Amino Acid Requirements in Human Nutrition: Report of a Joint FAO/WHO/UNU Expert Consultation*; WHO: Geneva, Switzerland, 2007.
42. Li, W.; Wen, L.; Xiong, S.; Xiao, S.; An, Y. Investigation of the effect of chemical composition of surimi and gelling temperature on the odor characteristics of surimi products based on gas chromatography-mass spectrometry/olfactometry. *Food Chem.* **2023**, *420*, 135977. [CrossRef]
43. Ahn, B.S.; Kim, B.G.; Jeon, E.B.; Lee, I.S.; Oh, K.S. Quality characteristics by grade of commercial frozen surimi. *Korean J. Fish. Aquat. Sci.* **2019**, *52*, 555–561.
44. KS—Korean Standard for Fish Cake (H 6017). In *Evaluation Criteria for KS Mark Certification*; Ministry of Oceans and Fisheries: Sejong, Republic of Korea, 2023. Available online: https://www.kats.go.kr/en/content.do?cmsid=406 (accessed on 11 November 2024).
45. Cha, S.H.; Jo, M.R.; Lee, J.S.; Lee, J.H.; Ko, J.Y.; Jeon, Y.J. Preparation and texture characterization of surimi gel using a unmarketable rearing Olive Flounder. *Korean J. Fish. Aquat. Sci.* **2009**, *42*, 109–115.
46. Yamazawa, M.; Murase, M.; Shiga, I. Improvement of the quality of retorted Kamaboko-I. Influence of heating conditions on the quality of Kamaboko. *Bull. Jpn. Soc. Sci. Fish.* **1979**, *45*, 187–192. [CrossRef]
47. Minh, N.P.; Nga, N.H. Different conditions impacting to physicochemical properties and sensory characteristics of bronze featherback sausage. *Int. J. Appl. Eng. Res.* **2018**, *13*, 1328–1331.
48. Kang, S.I.; Kim, J.S.; Park, S.Y.; Cho, H.J.; Jang, M.S.; Oh, J.Y.; Choi, J.S. Development and quality attributes of paste sausage supplemented with common squid (*Todarodes pacifics*) tailored for the elderly. *Appl. Sci.* **2023**, *13*, 10735. [CrossRef]
49. Embuscado, M.E. Spices and herbs: Natural sources of antioxidants—A mini review. *J. Funct. Foods* **2015**, *18*, 811–819. [CrossRef]
50. Park, J.H.; An, D.S.; Lee, D.S.; Park, E. Prediction of shelf-life of sea tangle drink. *J. Korean Soc. Food Sci. Nutr.* **2014**, *43*, 784–790. [CrossRef]
51. MFDS—Ministry of Food and Drug Safety. *6th General Analysis Method. Foodcode (Sik-Poom-Gong-Jeon)*; MFDS: Cheongju, Republic of Korea, 2023. Available online: https://various.foodsafetykorea.go.kr/fsd/#/ext/Document/FC (accessed on 31 August 2023).
52. Kim, H.J.; Yoon, M.S.; Park, K.H.; Shin, J.H.; Heu, M.S.; Kim, J.S. Processing optimization of gelatin from rockfish skin based on yield. *Fish. Aquat. Sci.* **2010**, *13*, 1–11. [CrossRef]
53. Bezerra, M.A.; Santelli, R.E.; Oliveira, E.P.; Villar, L.S.; Escaleira, L.A. Response surface methodology (RSM) as a tool for optimization in analytical chemistry. *Talanta* **2008**, *76*, 965–977. [CrossRef]
54. RDA—Rural Development Administration. *9th Revision Korean Food Composition Table*; Rural Development Administration: Jeonju, Republic of Korea, 2016.
55. Kang, S.I.; Kim, K.H.; Lee, J.K.; Kim, Y.J.; Park, S.J.; Kim, M.W.; Choi, B.D.; Kim, D.S.; Kim, J.S. Comparison of the food quality of freshwater rainbow trout *Oncorhhynchus mykisscultured* in different regions. *Korean J. Fish. Aquat. Sci.* **2014**, *47*, 103–113.
56. Park, J.W. *Surimi and Surimi Seafood*, 3rd ed.; Marcel Dekker Inc.: New York, NY, USA, 2013; p. 490.
57. Teixeira, B.; Marques, A.; Ramos, C.; Neng, N.R.; Nogueira, J.M.; Saraiva, J.A.; Nunes, M.L. Chemical composition and antibacterial and antioxidant properties of commercial essential oils. *Ind. Crops Prod.* **2013**, *43*, 587–595. [CrossRef]
58. Bligh, E.G.; Dyer, W.J. A lipid method of total lipid extraction and purification. *Can. J. Biochem.* **1959**, *37*, 911–917. [PubMed]
59. Bashir, K.M.I.; Park, Y.J.; An, J.H.; Choi, S.J.; Kim, J.H.; Baek, M.K.; Choi, J.S. Antioxidant properties of *Scomber japonicus* hydrolysates prepared by enzymatic hydrolysis. *J. Aquat. Food Prod. Technol.* **2018**, *27*, 107–121. [CrossRef]
60. Park, J.-R.; Kim, E.-G.; Jang, Y.-H.; Kim, K.-M. Screening and identification of genes affecting grain quality and spikelet fertility during high-temperature treatment in grain filling stage of rice. *BMC Plant Biol.* **2021**, *21*, 263. [CrossRef]

gels

MDPI

Article

Incorporation of Locust Bean Gum and Solid Lipid Microparticles as Strategies to Improve the Properties and Stability of Calcium-Rich Soy Protein Isolate Gels

Thais C. Brito-Oliveira, Ana Clara M. Cavini, Leticia S. Ferreira, Izabel C. F. Moraes and Samantha C. Pinho *

Laboratory of Encapsulation and Functional Foods (LEnAlis), Department of Food Engineering, School of Animal Science and Food Engineering (FZEA), University of São Paulo (USP), Pirassununga 13635-900, Brazil; thacbrito@gmail.com (T.C.B.-O.); leticia2.ferreira@usp.br (L.S.F.)

* Correspondence: samantha@usp.br; Tel.: +55-19-35654288

Abstract: The present study aimed to investigate the properties of calcium-rich soy protein isolate (SPI) gels (14% SPI; 100 mM $CaCl_2$), the effects of incorporating different concentrations locust bean gum (LBG) (0.1–0.3%, w/v) to the systems and the stability of the obtained gels. Also, the incorporation of solid lipid microparticles (SLMs) was tested as an alternative strategy to improve the system's stability and, therefore, potential to be applied as a product prototype. The gels were evaluated regarding their visual aspect, rheological properties, water-holding capacities (WHCs) and microstructural organizations. The $CaCl_2$-induced gels were self-supported but presented low WHC (40.0% ± 2.2) which was improved by LBG incorporation. The obtained mixed system, however, presented low stability, with high syneresis after 10 days of storage, due to microstructural compaction. The gels' stability was improved by SLM incorporation, which decreased the gelled matrices' compaction and syneresis for more than 20 days. Even though the rheological properties of the emulsion-filled gels (EFGs) were very altered due to the ageing process (which may affect the sensory perception of a future food originated from this EFG), the incorporation of SLMs increased the systems potential to be applied as a calcium-rich product prototype.

Keywords: soy protein isolate; salt-induced gels; calcium; plant proteins; locust bean gum; emulsion-filled gels

1. Introduction

The food industry is constantly facing challenges to adapt their formulations to a large variety of global changes and demands. Recently, the biggest issues of the sector are related to the global population increase (projected to achieve 9 billion by 2050), increase in life expectancy, concerns regarding the use of animal-based products and awareness of the importance of using sustainable ingredients [1,2]. In response to these tendencies, investigations regarding the functional properties of plant proteins have been growing significantly [3,4]. Most of the studies point out the challenges of applying plant proteins due to their distinct techno-functional properties in comparison to animal ingredients [5,6].

Soy protein isolate (SPI), for example, is the most important plant protein ingredient, known for its high availability in the market, high nutritional value and affordability, but also for presenting poor solubility, which affects many other functionalities, including its gelling capacity [7–11]. This fact was observed in a previous investigation by our research group, in which SPI was applied to produce NaCl-induced gels, and self-supported systems could be obtained only using relatively high protein concentrations (14%) [7]. Also, the strength and stability of the gels were not satisfactory, requiring the incorporation of galactomannans to improve the systems' properties [7]. According to the literature, the incorporation of polysaccharides is one of the most interesting strategies to improve the gelling capacities of protein ingredients, as they stabilize the microstructures formed [12–15]. In fact, the incorporation of locust bean gum in soy protein gels [7] was important to produce more stable structures which was later successfully applied to produce emulsion-filled gels encapsulating beta-carotene [16].

Even though NaCl-induced gelation methods have received attention, for allowing gelation at lower protein concentrations and lower temperatures than other methods [7,17,18], the consumption of sodium has been associated with a globally high prevalence of hypertension and cardiovascular disease, especially for elderly people [19]. One alternative to overcome such problems could be the replacement of NaCl by $CaCl_2$ in gel protein-based matrices.

In addition to allowing the reduction of sodium, such an alternative would increase the concentration of calcium in the potential food formulation, which is also a recent and important issue in the food industry [20,21]. According to the literature, calcium is an important raw material for bone formation, also playing important roles in other physiological activities, including coagulation, nerve and muscle activities and heart electrophysiology, among others. In this context, the maintenance of a stable extracellular calcium concentration may be seen as a homeostatic priority, especially for elderly people [20,21]. According to the literature, the World Health Organization (WHO) recommendations for calcium intake varies depending on the age, but for healthy adults is 1000 mg/day, and fortified foods can provide 30% of such recommended intake in 100 g [21].

The development of gelled food formulation with calcium and plant-based proteins is a quite complex process, requiring not only further investigations regarding the gelling capacity of SPIs and their behavior at high $CaCl_2$ concentrations, but a comprehension of the stability of such gels. The stability of gels is a very important research subject, due to the highly transient and non-linear nature of such systems, which tend to present structural alterations over time, due to spontaneous processes (i.e., aging processes) and/or external forces [22,23]. Such structural rearrangements can be either subtle or more drastic, even leading to the collapse of the gels, which, for food formulations, may affect the shelf life and the acceptability of the systems [22,24–26]. Considering this, the quantitative prediction of stability of gels is a critical for the formulation and manufacture of commercial food products [25].

Besides polysaccharides, the properties and stability of gelled systems may be affected by the addition of lipid fillers (e.g., emulsion or solid lipid microparticles) to the matrices, originating "emulsion filled gels" (EFGs). The properties of such complex systems depend on the characteristics of the gelled matrix, the lipid fillers and interactions between such components [27–29]. If it presents a positive effect over the gel's properties, the incorporation of such fillers may be interesting for future encapsulations of hydrophobic bioactive compounds, originating fortified food formulations.

Considering the promising advances of the last studies involving the cold-set gelation of soy protein isolate, but also the concerns regarding sodium consumption, the present study aimed to investigate: (i) the properties of $CaCl_2$-induced gels produced using the same SPI concentration (14%) and same ionic strength (μ: 100—representing 100 mM $CaCl_2$) as the ones previously applied for producing NaCl-induced gels [7], (ii) the effects of incorporating different concentrations of LBG to the systems and (iii) the stability of the obtained plant-based gel rich in calcium, and the effects of solid lipid microparticle incorporation to verify its potential application as a product prototype, for future incorporation of hydrophobic bioactives.

2. Results and Discussion

2.1. Production and Characterization of $CaCl_2$-Rich Cold-Set Gels: A Comparative Discussion

To verify the possibility of replacing NaCl by $CaCl_2$ as a gelling agent, the same SPI concentration (14%) and ionic strength (μ: 300) applied in our previous investigation using NaCl was tested with the divalent salt ($CaCl_2$). As verified in Figure 1I, the formulation tested resulted in the formation of an opaque self-supported gel, with an opaquer visual aspect in comparison to the systems previously produced using NaCl as a gelling agent [7]. According to the literature, the visual aspects of gelled systems are important to give some valuable information about the nature of protein aggregates, these being directly related to the type of matrix formed [30,31]. In the present investigation, for example, the opaque aspect of the gelled systems resulted from the particulate, disorganized, and porous microstructures, verified in the SEM micrograph Figure 1E, which generally result from a large and random aggregation process [30,31]. From Figure 1E, it was also possible to verify the formation of structures similar to "capsules" in a given region (highlighted in red). Such microstructural organization may be justified by three main factors: (I) the gelation method, involving direct salt addition; (II) the type of salt applied and (III) the mechanisms involved in the gelation of soy proteins.

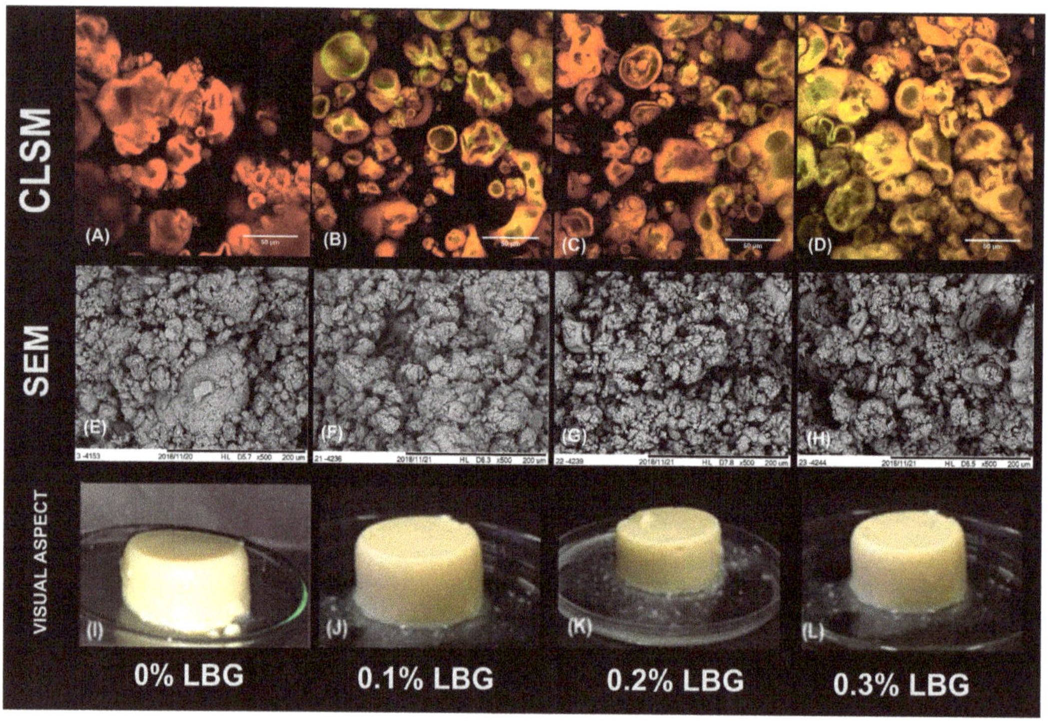

Figure 1. Confocal laser scanning micrographs (CLSMs) (in which Red identifies the protein phase; green identifies the polysaccharide phase) (**A–D**), scanning electron micrographs (SEMs) (**E–H**) and visual aspects (**I–L**) of gels produced with different locust bean gum (LBG) concentrations.

It is known that direct salt addition in cold-set methods may cause an excessively fast gelation process, resulting in random aggregations which generally lead to the formation of weaker matrices, with less ordered microstructures, such as the one shown in Figure 1E [32,33]. In terms of the type of salts, according to the literature divalent salts (i.e., $CaCl_2$) tend to cause much faster and effective interactions among protein aggregates in comparison to monovalent salts (i.e., NaCl), especially due to their higher capacity to screen electrical charges and to form cross-links among negatively charged carboxyl groups [7,30].

The formation of "capsule-like" structures in some regions of the gels, in contrast, possibly resulted from the aggregation processes that occurred during preheating of the SPI (rich in β-conglycinin and glycinin) and subsequent $CaCl_2$ addition. In their study, Guo et al. [34] observed that when β-conglycinin is heated in the presence of glycinin at pH 7 (i.e., the pH applied in the gels produced in the present investigation), β-conglycinin's hydrophilic groups tend to occupy the surface of the complex aggregates formed mainly by glycinin. In this context, it is possible that during the preheating, the aggregation process described by Guo et al. [34] led to an initial formation of the "capsule-like" structures verified in Figure 1E. Considering that the protein concentration used was below the critical concentration for the formation of heat-induced gels [7], such aggregates were still dispersed and only formed a gelled continuous matrix after the incorporation of salt. The microstructural organization observed resembles the ones referred to as microgels by some authors [35]. According to Nicolai and Durand [35], during primary protein aggregation, a finite size of aggregates is formed with, basically, two distinct morphologies: thin elongated aggregates and roughly spherical aggregates, which have also been called microgels (fractal, self-similar, branched aggregates or clusters), implying that the particle (in the present study referred as a "capsule") is formed by a crosslinked network and contains a relatively large fraction of solvent [35].

The stronger aspect of $CaCl_2$-induced gels was confirmed by the rheological data, presented in Figure 2 and Table 1. It was verified that NaCl-induced gels [7] were weaker than $CaCl_2$-induced ones (Table 1), presenting smaller values of G', G'', K' and K''. In addition, $CaCl_2$-induced gels presented higher n' and n''. Such behavior is related to the fact that, even though both monovalent and divalent salts are able to screen electrostatic interactions between charged protein molecules, resulting in decreases in electrostatic repulsions and protein aggregation, only the divalent cations can act as bridges between the negatively charged carboxylic groups on neighboring protein molecules, directly stabilizing the interactions between them [36,37]. Possibly due to the formation of these salt bridges, $CaCl_2$-induced gels presented G' and G'' with higher frequency dependence in comparison to the reported NaCl-induced gels, indicating a higher contribution of physical interactions in these systems [38]. Additionally, tan δ values were higher and more frequency-dependent for these $CaCl_2$-induced gels, indicating lower elastic contributions.

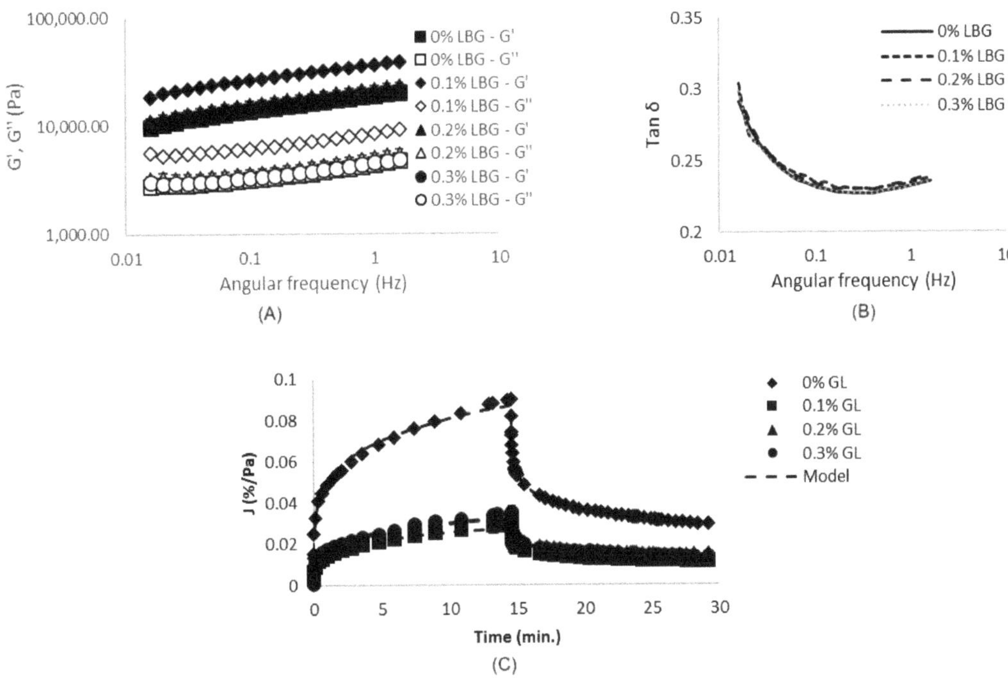

Figure 2. Results of frequency sweep (**A**,**B**) and creep/recovery tests (**C**) of gels produced with different locust bean gum (LBG) concentrations.

Table 1. Parameters of Power Law model (frequency sweep data), Burger's model (creep/recovery data) and uniaxial compression data of SPI gels produced with different locust bean gum (LBG) concentrations.

Analysis	Parameter	LBG Concentration			
		0%	0.1%	0.2%	0.3%
Frequency sweep	K′ (Pa)	18,881 ^{b} ± 1109	36,225 ^{a} ± 1750	21,322 ^{b} ± 1865	19,550 ^{b} ± 1072
	n′	0.150 ^{a} ± 0.003	0.151 ^{a} ± 0.002	0.149 ^{a} ± 0.002	0.146 ^{a} ± 0.002
	R^2	0.99	0.99	0.99	0.99
	K″ (Pa)	4233 ^{b} ± 248	8208 ^{a} ± 401	4852 ^{b} ± 433	4376 ^{b} ± 263
	n″	0.115 ^{a} ± 0.004	0.117 ^{a} ± 0.001	0.115 ^{a} ± 0.003	0.112 ^{a} ± 0.002
	R^2	0.96	0.96	0.95	0.95
Creep/recovery	J_0 (Pa^{-1})	0.031 ^{a} ± 0.001	0.009 ^{b} ± 0.002	0.010 ^{b} ± 0.001	0.011 ^{b} ± 0.002
	J_1 (Pa^{-1})	0.021 ^{a} ± 0.001	0.006 ^{b} ± 0.002	0.007 ^{b} ± 0.001	0.008 ^{b} ± 0.001
	λ_{ret} (s)	39.58 ^{a} ± 1.32	58.61 ^{a} ± 31.97	55.52 ^{a} ± 5.93	66.05 ^{a} ± 9.72
	η_0 (Pa.s)	27,865 ^{b} ± 1232	76,547 ^{a} ± 17,173	5995 ^{a} ± 2398	63,553 ^{a} ± 7003
	R^2	0.88	0.90	0.91	0.91
	ε (%)	68.0 ^{a} ± 1.9	61.7 ^{a} ± 8.6	56.3 ^{a} ± 1.9	62.1 ^{a} ± 2.8
Uniaxial compression	E_g (Pa)	4469 ^{a} ± 97	4064 ^{b} ± 82	2877 ^{c} ± 248	2868 ^{c} ± 102
	σ_H (Pa)	1301 ^{a} ± 333.9	1519 ^{a} ± 199	845 ^{b} ± 82	859 ^{b} ± 93
	ε_H	0.288 ^{a} ± 0.046	0.346 ^{a} ± 0.056	0.212 ^{b} ± 0.031	0.225 ^{b} ± 0.014

Averages followed by different lowercase letters are statistically different ($p < 0.05$) for gels with different concentrations of LBG.

The higher strength and lower elasticity of $CaCl_2$-induced gels in comparison to NaCl-induced ones may be explained by the microstructural organization of the systems produced with the divalent salt, which revealed to be much more particulate, as previously verified and discussed from Figure 1E [7].

Such conclusions were confirmed by the results of creep/recovery tests (Figure 2C), which indicated that $CaCl_2$-induced gels presented higher hardness and lower elasticities and abilities to recover after stress removal in comparison to NaCl-induced gels [7]. It is possible that, for $CaCl_2$-induced gels, the stress application caused disruptions in the salt bridges initially formed between the charged carboxylic groups, reducing the systems' ability to recover when the stress was removed, in comparison to NaCl-induced systems.

The uniaxial compression tests, on the other hand, showed that $CaCl_2$-induced gels presented a higher Young's modulus and rupture stress in comparison to the reported data for NaCl-induced gels [7]. Rupture strain, however, was less affected, as significant differences ($p < 0.05$) were only verified for systems with 14% SPI at ionic strength of 300 (100 mM $CaCl_2$), in which the incorporation of $CaCl_2$ resulted in the formation of structures with higher rupture strains.

Even though the obtained gel presented a self-supported aspect, the system exhibits a low water-holding capacity (40.0% ± 2.2). In general, particulate and porous gelled matrices, such as the ones observed in the present study, tend to present low WHC, especially for the lower capacities to retain water by capillary forces [39]. Besides it is known that divalent cations tend to form bridges among between biopolymers chains and, therefore, decrease the amount of hydrophilic groups available to interact with water, which remains free in the pores of the network.

2.2. Effects of the Incorporation of Different Locust Bean Gum in $CaCl_2$-Induced Gels

The low water-holding capacity is a potential problem in the development of food gels and highlights the challenging properties of the SPI. As previously cited, however, the incorporation of polysaccharides is a potential alternative to improve the gelling capacity of protein ingredients [7,40]. In this context, different LBG concentrations were incorporated to $CaCl_2$-induced gels and the systems were characterized, initially in terms of its microstructural organizations (Figure 1F–H). It was verified that the higher the amount of LBG, the higher the heterogeneity of the systems. The increase in polysaccharide concentration up to 0.2% led to the formation of smaller and more separated aggregates. Systems containing 0.3% LBG, however, presented some regions with a more compact aspect, mainly formed by microgel-like structures.

CLSM micrographs (Figure 1B–D) revealed that the formulations with 0.1% and 0.2% LBG presented a certain degree of phase separation, with some reddish and others greenish regions, while systems containing 0.3% LBG presented a lower extension of such phenomenon. According to the literature, thermodynamic incompatibility and phase separation are very commonly observed phenomena in aqueous mixtures of protein and neutral polysaccharides. However, when gelation occurs, the process of phase separation tends to slow down or eventually stop, with the resulting microstructures being "arrested" within the formed polymer network, as verified from the incomplete demixing processes verified, especially at higher LBG concentrations [40].

As expected, such microstructural organizations reflected in the rheological properties of the systems, shown in Figure 2 and Table 1. The incorporation of 0.1% LBG resulted in the strengthening of the matrices, as verified through the increases in G' and G'' (and, therefore, K' and K''). The addition of higher LBG concentrations, however, did not significantly change ($p > 0.05$) the values of K' and K'' in comparison to SPI only gels. For all formulations, on the other hand, the values of tan δ, n' and n'' were similar.

The results of the creep/recovery tests (Figure 2C), however, indicated a decrease in compliance values for all three LBG concentrations tested. Such results revealed that, even though systems with 0.1% LBG presented a higher strength under small deformations, the presence of such galactomannan increased the resistance of the gels to deform under the application of the constant stress during the creep step, resulting in decreases in J_0 and J_1, in comparison to SPI-only gels. In addition, LBG-SPI mixed gels presented higher η_0. λ_{ret} and the recovery capacities of the systems, on the other hand, were not significantly affected.

The strengthening verified through the incorporation of LBG (especially at 0.1%) is related to the increase in protein–protein interactions (mainly through saline bridges between charged carboxylic groups) and the higher self-association of polysaccharides, which possibly reduced the effects of incomplete demixing, and resulted in the more evident microphase separations previously discussed.

Uniaxial compression data (Table 1), on the other hand, revealed that the LBG incorporation decreased E_g in all concentrations and decreased the rupture parameters in concentrations equal or higher than 0.2%. Therefore, even though LBG incorporation increased the strength of the gels under oscilatory/small deformations, it compromises system strength under large deformations, possibly due to the heterogeneous and more porous microstructures previously discussed. According to the literature, even though microphase separations may increase interactions among biopolymers within each phase, and also the elastic response of systems, such effect depends on the connectivity between protein aggregates [7,41]. Therefore, it was possibly the fragile connectivity within the systems with higher LBG concentrations (0.2 and 0.3%) resulted in the lowest rupture parameters verified. Such fragility, however, did not compromise the strength of the systems under smaller deformations, as verified from frequency sweep tests and creep/recovery tests.

Regarding the water holding-capacity, systems with 0.1% LBG presented the higher values, with an average of 46.5% ± 0.4, followed by 0.2% LBG (WHC: 45.0% ± 1.5) and 0.3% LBG (42.1% ± 2.2). Observing the results obtained in this section, it is clear that the formulation with 0.1% LBG was the most promising, with a relatively higher degree of phase separation, higher WHC and higher resistance in terms of rupture properties under large deformations and general higher resistance under small deformations. Therefore, such formulation was selected to the next step of this investigation, which consisted in evaluating the stability of the gels.

2.3. Evaluation of Gel Stability

The quantitative prediction of gelled systems' stability is considered a critical issue in the formulation of commercial food products [25], for this reason the formulation produced with using SPI (14%), LBG (0.1%) and $CaCl_2$ (100 mM) was evaluated during storage. For the first day, all the characterization data previously discussed were considered and, therefore, the gels were self-supported and presented an opaque aspect (Figure 3A). Even though the system presented a relatively low WHC (46.5% $\pm$ 0.4), the gel could be adequately analyzed through frequency sweep tests, creep/recovery tests and uniaxial compression tests, as previously shown in Figure 2 and Table 1. After 10 days, however, the excessive syneresis verified in Figure 1C prevented the development of rheological analyzes for compromising the accommodation of the samples in the equipment. For this reason, at the 10th day of storage, the gels were only analyzed regarding the visual aspect (Figure 1C) and microstructural organization (Figure 1D). The micrograph evidenced that the aging process led to a significant compaction in the matrices' microstructures, which led to the expulsion of the water initially trapped in the gels' pores (extensive syneresis) and prevented further characterizations in subsequent days. According to Alting et al. [38], open clusters of aggregates (similar to the ones verified at day 01) are thermodynamically unstable and can be partly stabilized by the formation of additional bonds, which causes the compaction verified at the 10th day of storage (Figure 1D).

Figure 3. SEM micrographs and visual aspects of non-filled gels at days 1 (**A**,**C**) and 10 (**B**,**D**) of storage.

According to Barlett et al. [25], a gel is a metastable phase with a high free-energy density whose consolidation is driven by a force for phase separation. When the network is formed, however, the dynamics of phase separation are slowed down. With ageing, the network lowers its free energy via structural reorganizations which proceed through the rupture of single-particle bonds, diffusion to denser region of the network, and a reformation of broken bonds, with a net increase in the number of nearest-neighboring particles [25]. Such processes tend to make the network coarsen, as verified in the present investigation.

Even though the tendency of the gelled systems is to suffer microstructural rearrangements to low free energy, it is known that the velocity through which these alterations will occur up to a point of compromising a food products acceptability can be altered by formulation changes. For this reason, solid lipid microparticles were incorporated to the gels as a possible strategy to improve their stability, decreasing the undesirable changes, which led to the excessive water loss from the matrices, and also for opening a possibility for future bioactive incorporations to the formulation.

2.4. Incorporation of Solid Lipid Microparticles as a Strategy to Improve the Gels' Stability/Shelf-Life

Solid lipid microparticles were, then, incorporated to the gels and the obtained emulsion-filled gels (EFGs) were initially characterized in terms of their visual aspect, microstructural organization and water-holding capacity on the first day of storage, and the obtained data are shown in Figure 4. The EFGs were self-supported gels (Figure 4A), with larger aggregates/structures which resembled capsules or microgels (Figure 4B), in comparison to the non-filled systems. As verified in Figure 1C,D, the presence of SLMs decreased the extent of incomplete demixing, increasing the microphase separations in the micrographs. Considering that the microstructural organizations of such mixed gels result from the competition between gelation and the phase separation process [40,41], and also that the salt concentration and incorporation methods were the same for non-filled gels and EFG, it is hardly possible that the SLMs affected the gelation rate. However, the presence of SLMs probably increased the rate of phase separation, due to the low chemical affinity between the protein ingredient and the surfactants Tween 80 and Span 80, which increased the thermodynamical incompatibility within the gels and accelerated the phase separation process [28,29].

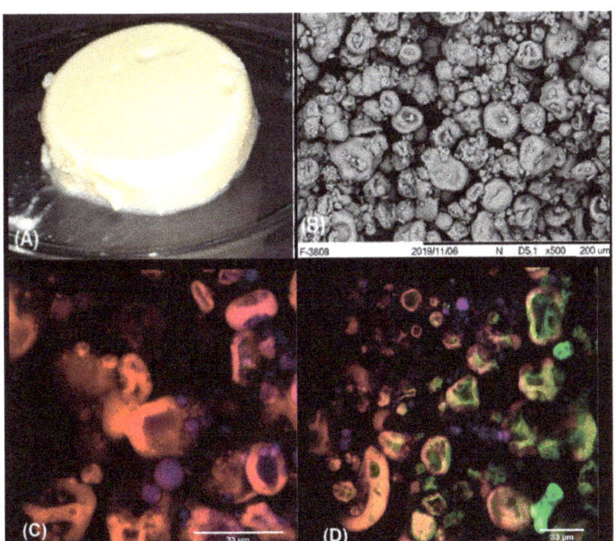

Figure 4. Visual aspect (**A**), scanning electron micrographs (SEMs) (**B**) and confocal laser scanning micrographs (CLSMs) (**C**,**D**) of emulsion-filled gels at the first day of storage.

In some studies found in the literature, the authors affirm that phase separation in protein–polysaccharide mixed gels may increase local biopolymer concentrations in each phase, which favors a more extensive polymer self-association, with potential increases in the gels' strength [40]. In the present investigation, however, such an effect was not verified, as shown in Table 2. The results demonstrated that the incorporation of SLMs to the gels decreased their strength, as verified from the lower values of E_g and rupture parameters of the EFGs in comparison to non-filled gels.

Table 2. Parameters of Power Law model (frequency sweep data), Burger's model (creep/recovery data) and uniaxial compression data of emulsion-filled gels (EFGs) at different days of storage.

Formulation	Parameter	Day 01	Day 10	Day 20
Water-holding capacity	WHC (%)	65.0 ^{a} ± 4.6	61.7 ^{a} ± 3.1	63.3 ^{a} ± 0.5
Frequency sweep tests	K' (Pa)	8103 ^{c} ± 318	10,678 ^{b} ± 569	2011 ^{a} ± 562
	n'	0.1367 ^{a} ± 0.0023	0.1238 ^{b} ± 0.0018	0.1229 ^{b} ± 0.0038
	R^2	0.99	0.99	0.99
	K'' (Pa)	1698 ^{c} ± 50	2192 ^{b} ± 131	4060 ^{a} ± 92
	n''	0.0754 ^{a} ± 0.0042	0.0710 ^{a} ± 0.0074	0.0805 ^{a} ± 0.0035
	R^2	0.86	0.82	0.91
Creep/recovery tests	J_0 (%/Pa)	0.0255 ^{a} ± 0.0022	0.0123 ^{b} ± 0.0003	0.0109 ^{b} ± 0.0017
	J_1 (%/Pa)	0.0231 ^{a} ± 0.0013	0.0084 ^{b} ± 0.0009	0.0101 ^{b} ± 0.0021
	λ_{ret} (s)	99.06 ^{ab} ± 5.21	89.26 ^{b} ± 9.10	107.73 ^{a} ± 4.67
	η_0 (Pa.s/%)	30,914 ^{b} ± 4674	34,352 ^{b} ± 2886	110,475 ^{a} ± 29,707
	R^2	0.91	0.93	0.89
	ε (%)	65.7 ^{a} ± 4.8	42.4 ^{b} ± 5.8	78.0 ^{a} ± 11.6
Uniaxial compression tests	E_g (Pa)	3161 ^{b} ± 419	7533 ^{a} ± 436	7857 ^{a} ± 537
	σ_H (Pa)	1092 ^{b} ± 72	2115 ^{a} ± 209	2348 ^{a} ± 240
	ε_H	0.2398 ^{b} ± 0.0097	0.2751 ^{ab} ± 0.0220	0.2956 ^{a} ± 0.0191

Averages followed by different lowercase letters in the same line are statistically different ($p < 0.05$) for EFGs at different days of storage.

In this case, the rheological data may also be explained by their microstructural organization. Possibly, before the gelation process, the molecules of proteins and polysaccharides were confined in some regions of the samples as a result of the repulsive interactions between the Tween 80/Span 80-stabilized SLMs and biopolymers. With $CaCl_2$-incorporation, however, an extensive polymer self-association took place in punctual regions of the systems (in which the biopolymers were concentrated), leading to the formation of the "capsules/microgel"-like structures. Although the microphase separation was more intense within the microgels, the connections among these individual structures were potentially compromised by the presence of the SLMs (located in the spaces among the microgels as shown in Figure 4C,D), which made the matrix more heterogeneous, discontinuous and, consequently, weak.

On the other hand, EFGs presented higher WHC in comparison to non-filled gels (Table 2), also due to the microstructural organization composed by the microgels. Considering that microgels are generally formed by a crosslinked network and contain a relatively large fraction of solvent inside, it is possible that the increase in water retention in EFGs was more related to the amount of liquid entrapped within the microgels than in the spaces among them [35]. Otherwise, the variations in the spaces among these structures would have caused higher WHC variations, due to distinct capillarity forces.

After this initial characterization, the EFGs were stored and evaluated during a 20-day storage period and the results are shown in Figure 5. The visual aspects of the gels on the 10th and 20th day of storage evidenced that the presence of SLMs decreased the loss of water from the gelled systems during storage. Noticeably, the EFGs did not present syneresis during the 20 days analyzed. After this period, however, the EFGs started to lose water and presented microbial growth, which limited the analyzes in subsequent days of storage.

Figure 5. Visual aspects and SEM micrographs of EFGs at different days of storage.

The cited reduction in water loss in the EFGs is related to the microstructural organizations, also shown in Figure 5. Even though the ageing process of the EFGs also led to a microstructural compaction, it was much more subtle than that observed in non-filled gels, indicating that the free energy of the systems containing the SLMs (mainly formed by microgels) was lower than that of particulate non-filled gels. Considering the forms and sizes of microgels were very similar during storage, only with different proximities, it is possible that the ageing process only involved the formation of new bonds among the microgels, and not the breakup of existing bonds within the structural units.

The higher stability of EFGs was also confirmed through the WHC values, which remained constant during the 20 days analyzed (Table 2). Such data reinforced the fact that the larger amount of liquid in the EFGs was entrapped inside the microgels and not among them, not being largely affected by the compaction of the systems, which brought the microgels closer to each other.

Such higher WHC values resulted in a better accommodation of the EFGs in the rheometer allowing the characterization of such systems in the different days of storage. The obtained results are shown in Figure 6 and Table 2. Frequency sweep tests revealed a continuous increase in G′, G″, and, consequently, K′ and K″, and a decrease in n′ during the 20 days of storage. Such strengthening of the EFGs revealed that, for such systems, the ageing process led to the formation of more stable bonds between protein groups reducing the free energy and increasing the microstructural compaction. In addition, it was verified a continuous decrease in the values of tan δ (Figure 6B), indicating an increase in the elastic responses of the EFGs, possibly due to the more effective interactions of the biopolymers after the compaction of the microstructures.

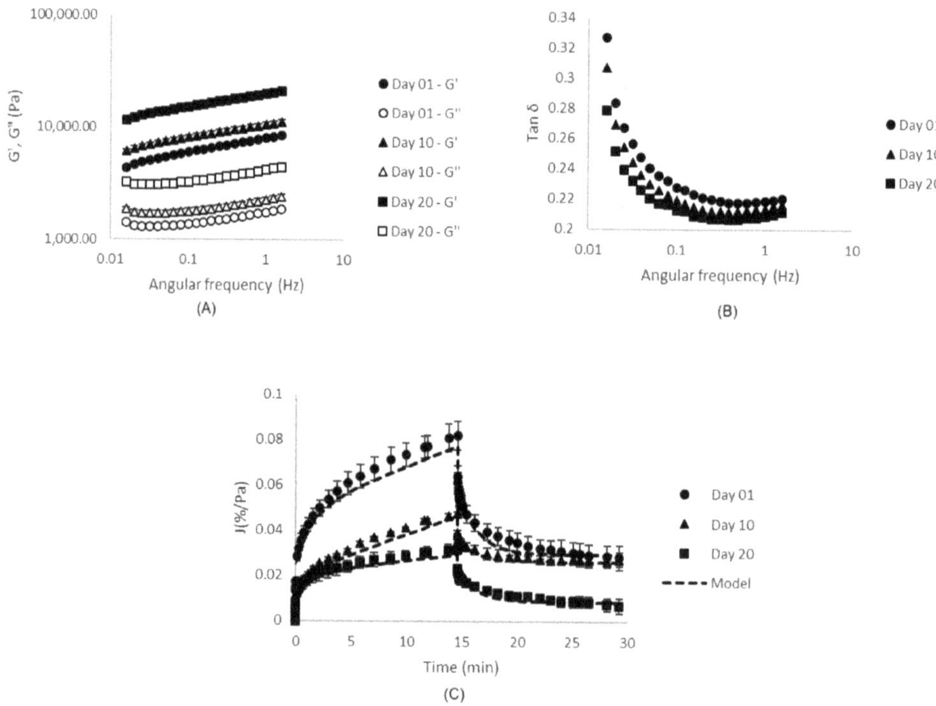

Figure 6. Results of frequency sweep tests (**A**,**B**) and creep/recovery tests (**C**) of EFGs on different days of storage.

Such alterations also led to decreases in compliance (Figure 6C), J_0 and J_1, and increases in η_0 (Table 2). Interestingly, on the 10th day of storage, the systems had the lowest values of λ_{ret} and recovery capacity, indicating that, although new connections and interactions were established, they were still fragile. On the 20th day, however, such interactions had become more stable, resulting in the further increases in λ_{ret} and recovery capacity.

The strengthening (in this case, increased rigidity) of the EFGs during storage was confirmed by uniaxial compression data, which showed increases in all parameters after the 10th day of storage. Such results confirmed that the incorporation of SLMs was beneficial to the gels' stability as it decreased the microstructural alterations; however, the rheological properties of the gels were significantly altered. The influences of these changes in the sensory perception of a future food originating from these EFGs is difficult to predict. However, these data allow us to affirm that the incorporation of SLMs was an important step towards the reduction of water loss from the non-filled gels.

3. Conclusions

$CaCl_2$-induced gels presented particulate, disorganized, and porous matrices, which led to low WHC values. The obtained systems were stronger (higher G′, G″, Eg and rupture properties, and lower compliance values) in comparison to NaCl-induced gels produced in a previous investigation. The low WHC was improved with LBG incorporation, especially in concentrations of 0.1%. In such a concentration, the polysaccharide incorporation increased G′ and G″, indicating a strengthening of the matrices, but decreased E_g, due to the heterogeneous and more porous microstructures, with incomplete demixing. Gels with SPI (14%), LBG (0.1%) and $CaCl_2$ (100 mM) had their properties evaluated during storage, and presented very poor stability, as after 10 days the systems could not be adequately analyzed using a rheometer due to the excessive syneresis resulting from a significant microstructural compaction. The incorporation of solid lipid microparticles to such gels, however, was revealed to be an interesting strategy to improve the stability of the obtained plant-based gel rich in calcium, especially by reducing the microstructural compaction and, therefore, syneresis, for more than 20 days. Even though the rheological properties of the EFGs were significantly altered due to the ageing process (which may affect the sensory perception of a future food originated from such EFGs), the incorporation of solid lipid microparticles increased the systems' potential to be applied as a product prototype, for even for future incorporations of hydrophobic bioactives.

4. Materials and Methods

4.1. Chemicals and Reagents

For gel production, the SPI (Protimarti M-90, 84.3% protein) was purchased from Marsul (Montenegro, RS, Brazil), calcium chloride dihydrate from Kyma (Americana, SP, Brazil), and locust bean gum (Viscogum LBG®) has been donated by Cargill (Mairinque, SP, Brazil). For SLM production, palm stearin (PS), with melting point of 50.1 °C was donated by Agropalma (Belém, PA, Brazil), while Tween 80 and Span 80 were purchased from Sigma-Aldrich (St. Louis, MO, USA). In all experiments, the water applied was Ultrapure obtained from a Millipore system Direct Q3® (Billerica, MA, USA).

4.2. Production and Characterization of Cold-Set $CaCl_2$-Rich Gels

The production of cold-set SPI gels was performed using a protocol defined in a previous study [7]. For this purpose, a dispersion of 14% (*w*/*v*) SPI, with pH 7, was produced using deionized water at room temperature. Afterwards, the samples were preheated in a water bath at 80 °C for 30 min and cooled to room temperature. Subsequently, the systems were added with 100 mM $CaCl_2$ (i.e., 11.098 mg/mL). Such a salt concentration was calculated to achieve the same ionic strength previously applied to producing NaCl-induced gels [7]. The samples were, then, stored at 10 °C for 12 h before the characterizations.

For the production of SPI-LBG mixed gels, the same protocol was applied, however, but the locust bean gum (LBG) (0.1–0.3%, *w*/*v*) was added to the SPI powder before hydration. The different formulations were characterized in terms of visual aspects, water-holding capacities, microstructural properties (confocal laser scanning microscopy and scanning electron microscopy) and rheological parameters (uniaxial compression tests, small-amplitude oscillatory shear tests and creep/recovery tests).

4.3. Production of Emulsion-Filled Gels

4.3.1. Production of Solid Lipid Microparticles (SLMs)

The production of SLMs was performed according to a protocol described by Brito-Oliveira et al. [28]. For this purpose, the lipid phase (composed by palm stearin—4.5%, *w*/*w*, and Span 80—2.7%, *w*/*w*), was heated to 80 °C for 30 min to eliminate the thermal memory. Afterwards, the aqueous phase (composed by Tween 80—1.8%, *w*/*v*—and deionized water) was dispersed in the lipid phase, using a rotor–stator device (IKA T25, IKA, Staufen, Germany) at 18,000 rpm for 5 min, at 80 °C. The systems were added with sodium benzoate (0.02%, *w*/*w*) to avoid microbiological contamination. The solid lipid microparticles dispersions (SLMDs) were stored at 10 °C overnight before their application to produce EFGs.

4.3.2. Production of Emulsion-Filled Gels

For EFG production, the same protocol described in item 2.2 was applied, using a formulation of 14% (*w*/*v*) of SPI, 0.1% (*w*/*v*) of LBG and 100 mM of $CaCl_2$; however, 50% of the deionized water used to hydrate the SPI/LBGs was replaced by solid lipid microparticle (SLM) dispersions. The SLM concentration was selected after a preliminary scanning and observation of the visual aspect of the obtained gels. The systems were also stored at 10 °C for 12 h before characterization. The systems were characterized in terms of visual aspects, water-holding capacities, microstructural properties (confocal laser scanning microscopy and scanning electron microscopy) and rheological parameters (uniaxial compression tests, small-strain oscillatory tests and creep/recovery tests).

4.4. Evaluation of Gel Stability

For the evaluation of the gels' stability, non-filled gels and emulsion-filled gels went through evaluations of visual aspects, WHC, morphology (by scanning electron microscopy) and rheological parameters (by uniaxial compression, small-amplitude oscillatory shear and creep/recovery tests) in different days of storage.

4.5. Confocal Laser Scanning Microscopy

Confocal laser scanning microscopy (CLSM) (Confocal Upright Microscope LSM 780 NLO-Zeiss, Zeiss, Germany) was conducted according to protocol described by Brito-Oliveira et al. [29], using simultaneous dual-channel imaging. The protein phase was visualized by exciting rhodamine B (stock solution of 0.2%, *w*/*v*, in deionized water—application of 10 μL of stock solution/mL of gel) at a wavelength of 543 nm and at an emission wavelength range of 551–655 nm. The polysaccharides were visualized by exciting the dye fluorescein isothiocyanate (FITC) (stock solution of 1 mg/mL in dimethyl sulfoxide—application of 0.05 mL of stock solution/mL of gel) at an excitation wavelength of 488 nm and an emission wavelength range of 493–543 nm. The SLMs were visualized by exciting the Nile Red dye (stock solution of 0.1 g/100 mL in methanol—application of 10 μL of solution/g of lipid) at an excitation wavelength of 488 nm and an emission wavelength range of 500–580 nm.

4.6. Scanning Electron Microscopy

The scanning electron microscopy (SEM) was performed according to a protocol described by Picone, Takeuchi and Cunha [42]. For sample preparation, the gels (1.0 × 0.5 × 0.5 cm) were fixed in 2.5 g/100 g glutaraldehyde in a cacodylate buffer (16 g/L) at pH 7.2 and stored at 7 °C for 24 h. Subsequently, the systems were rinsed twice in a cacodylate sodium buffer (16 g/L, pH 7.2) and fractured in liquid nitrogen. Then, the samples were subjected to post-fixation using osmium tetroxide 1 g/100 g in a cacodylate buffer (16 g/L, pH 7.2) for 120 min and rinsed twice in deionized water. Afterwards, the gels were dehydrated in a graded ethanol series (30, 50, 70 and 90 mL/100 mL) for 20 min in each, and in 100% ethanol (three changes in 1 h). The dehydration was completed by critical-point drying (CPD03 Balzers Critical Point Dryer, Alzenau, Germany). The samples were, then, fractured, placed in aluminum stubs, coated with gold (200 s/40 mA) in a Balzers SCD 050 Sputter Coater (Alzenau, Germany), and analyzed using a TM 3000 tabletop microscope (Hitachi, Tokyo, Japan).

4.7. Water-Holding Capacity

The water holding capacity (WHC) of the gels was analyzed using protocol described by Beuschel, Culbertson, Partridge and Smith [43]. For this purpose, the systems were weighted on Whatman paper No. 1 placed in Falcon tubes and centrifuged at 2500 rpm for 10 min at 6 °C (Hermle centrifuge Labortechnik GmbH, model Z-216 MK, Wehingen, Germany). Subsequently, the samples were removed from the filters, which were weighted for the determination of the water mass released. The values of WHC were calculated using Equation (1).

$$\mathrm{WHC}(\%) = \left(1 - \left(\frac{m_f - \mathrm{m_i}}{\mathrm{m_s}}\right)\right) \times 100 \tag{1}$$

where m_f is the weight of the wet filter (g), $\mathrm{m_i}$ is the initial weight of the dry filter (g), and m_s is the weight of the SPI sample (1–2 g). Each experiment was performed in triplicate.

4.8. Uniaxial Compression

For the development of uniaxial compression tests, a protocol adapted from Oliver, Scholten and van Aken [44] was applied, using a texturometer (TA-XT.plus Texture Analyser, Godalming, Surrey, UK). For this purpose, the gels with a cylindrical shape were compressed to 80% of their original height using an aluminum probe lubricated with silicone oil to minimize friction. The formulations were tested using five replicates and a deformation speed of 1 mm/s. Hencky stress (σ_H) and Hencky strain (ε_H) values were calculated from the force–deformation data using Equations (2) and (3), respectively:

$$\sigma_H = \mathrm{F(t)} \cdot \frac{H(t)}{H_0 \cdot A_0} \tag{2}$$

$$\varepsilon_H = \ln \frac{H(t)}{H_0} \tag{3}$$

where F(t) is the force at time t, A_0 is the initial area, H_0 is the initial height, and $H(t)$ is the height at time t. The rupture parameters were associated with the maximum value of the stress–strain curve. The values of the apparent Young's modulus ($\mathrm{E_g}$) of the systems were determined by the slope of the first linear interval in the Hencky stress (σ_H) versus Hencky strain (ε_H) curves, up to rupture.

4.9. Small Strain Oscillatory Tests

Small amplitude oscillatory shear tests were developed using protocol adapted from Chang, Li, Wang, Bi and Adhikari [45]. The tests were performed in an AR2000 rheometer (TA Instruments, New Castle, DE, USA), at 10 °C, with aluminum parallel plate geometry (60 mm diameter, 1 mm gap). Silicone oil was used at the edge of the samples to avoid water evaporation, and a 2 min resting period was applied before each experiment. The linear viscoelastic region (LVR) of each sample was determined through strain sweep tests, which were developed in the range 0.01–100%, using a constant frequency of 1 Hz. Frequency sweep tests, on the other hand, were carried out over an angular frequency range of 0.016–1.6 Hz, using a strain amplitude of 2% (determined for being within the LVR). The angular frequency dependence of both viscoelastic moduli (G' and G'') was described by a power law model (Equations (4) and (5)), using the nonlinear regression feature in Excel (Microsoft, Seattle, WA, USA) [45]:

$$G' = K' \cdot \omega^{n'} \tag{4}$$

$$G'' = K'' \cdot \omega^{n''} \tag{5}$$

where K' and K'' are power law parameters, n' and n'' are frequency exponents, and ω is the angular frequency.

4.10. Creep/Recovery Tests

For creep/recovery tests, a protocol described by Brito-Oliveira et al. [7] was applied and the experiments were developed in an AR2000 rheometer (TA Instruments, New Castle, PN, EUA), using an aluminum parallel plate geometry (60 mm diameter, 2 mm gap), at 10 °C. A resting time of 10 min was applied for the elimination of loading effects. For the creep step, a constant stress of 5 Pa was applied for 15 min and then removed for the evaluation of the recovery behavior of the samples for another 15 min. Silicone oil was also used on the edges of the samples to avoid water evaporation. Creep and recovery data were analyzed using the four-parameter Burger's model represented by Equation (6) [45].

$$J(t) = \begin{cases} J_0 + J_1\left(1 - \exp\left(\frac{-t}{\lambda_{ret}}\right)\right) + \frac{t}{\eta_0},\ t \leq t_1 \\ J_1\left(\exp\left(\frac{t_1 - t}{\lambda_{ret}}\right) - \exp\left(\frac{-t}{\lambda_{ret}}\right)\right) + \frac{t_1}{\eta_0},\ t > t_1 \end{cases} \quad (6)$$

where J_0 is the instantaneous compliance in %/Pa, η_0 is the viscosity of the Maxwell dashpot in Pa·s/%, J_1 is the compliance associated with the Kelvin–Voigt element in %/Pa, λ_{ret} is the retardation time associated with the Kelvin–Voigt element in s, and t_1 is the time when the stress was removed.

The recovery rates were calculated using Equation (7), where ε_{max} is the maximum strain at the end of the creep test and ε_f is the final strain [45].

$$Recovery\ (\%) = \left(\frac{\varepsilon_{max} - \varepsilon_f}{\varepsilon_{max}}\right) * 100 \quad (7)$$

4.11. Statistical Analyses

All measurements were performed at least in triplicate, and mean values and corresponding errors were calculated. For the statistical treatment of data, an analysis of variance (ANOVA) was conducted, followed by Tukey's tests with a 5% significance level using SAS software version 9.2 (SAS Institute Inc, Cary, NC, USA).

Author Contributions: Conceptualization, T.C.B.-O. and S.C.P.; methodology, T.C.B.-O., I.C.F.M. and S.C.P.; validation, T.C.B.-O. and S.C.P.; formal analysis, T.C.B.-O., A.C.M.C. and L.S.F.; investigation, T.C.B.-O., A.C.M.C. and L.S.F.; data curation, T.C.B.-O., I.C.F.M. and S.C.P.; writing—original draft preparation, T.C.B.-O.; writing—review and editing, I.C.F.M. and S.C.P.; supervision, S.C.P.; project administration, S.C.P.; funding acquisition, T.C.B.-O. and S.C.P. All authors have read and agreed to the published version of the manuscript.

Funding: This research was funded by São Paulo Research Foundation (Fundação de Amparo à Pesquisa do Estado de São Paulo—FAPESP) grant number 2017/06224-9.

Data Availability Statement: The raw data supporting the conclusions of this article will be made available by the authors on request.

Acknowledgments: The authors thank the São Paulo Research Foundation (Fundação de Amparo à Pesquisa do Estado de São Paulo—FAPESP) for the Ph.D. fellowship provided to Thais C. Brito-Oliveira (grant number 2017/06224-9). We also thank the Electron Microscope Laboratory (LME/UNICAMP) for access to equipment and the assistance provided.

Conflicts of Interest: The authors declare that they have no conflicts of interest.

References

1. Willett, W.; Rockström, J.; Loken, B.; Springmann, M.; Lang, T.; Vermeulen, S.; Garnett, T.; Tilman, D.; DeClerck, F.; Wood, A.; et al. Food in the Anthropocene: The EAT–Lancet Commission on healthy diets from sustainable food systems. *Lancet* **2019**, *393*, 447–492. [CrossRef] [PubMed]
2. Pang, S.; Chen, M.C. Investigating the impact of consumer environmental consciousness on food supply chain: The case of plant-based meat alternatives. *Technol. Forecast. Soc.* **2024**, *201*, 123190. [CrossRef]
3. Akharume, F.U.; Aluko, R.E.; Adedeji, A.A. Modification of plant proteins for improved functionality: A review. *Compr. Rev. Food Sci. Food Saf.* **2021**, *20*, 198–224.
4. Ma, K.K.; Greis, M.; Lu, J.; Nolden, A.A.; McClements, D.J.; Kinchla, A.J. Functional performance of plant proteins. *Foods* **2022**, *11*, 594. [CrossRef] [PubMed]
5. Gao, K.; Rao, J.; Chen, B. Plant protein solubility: A challenge or insurmountable obstacle. *Adv. Colloid. Interface Sci.* **2024**, *324*, 103074. [CrossRef] [PubMed]
6. Sim, S.Y.J.; Srv, A.; Chiang, J.H.; Henry, C.J. Plant proteins for future foods: A roadmap. *Foods* **2021**, *10*, 1967. [CrossRef] [PubMed]
7. Brito-Oliveira, T.C.; Cavini, A.C.M.; Ferreira, L.S.; Moraes, I.C.; Pinho, S.C. Microstructural and rheological characterization of NaCl-induced gels of soy protein isolate and the effects of incorporating different galactomannans. *Food Struct.* **2020**, *26*, 100158. [CrossRef]

8. Ingrassia, R.; Bea, L.L.; Hidalgo, M.E.; Risso, P.H. Microstructural and textural characteristics of soy protein isolate and tara gum cold-set gels. *LWT* **2019**, *113*, 108286. [CrossRef]
9. Ningtyas, D.W.; Tam, B.; Bhandari, B.; Prakash, S. Effect of different types and concentrations of fat on the physico-chemical properties of soy protein isolate gel. *Food Hydrocoll.* **2021**, *111*, 106226. [CrossRef]
10. Ma, Z.; Li, L.; Wu, C.; Huang, Y.; Teng, F.; Li, Y. Effects of combined enzymatic and ultrasonic treatments on the structure and gel properties of soybean protein isolate. *LWT* **2022**, *158*, 113123. [CrossRef]
11. Xing, H.; Liu, X.; Hu, Y.; Hu, K.; Chen, J. Effect of Lycium barbarum polysaccharides on heat-induced gelation of soy protein isolate. *Food Hydrocoll.* **2024**, *147*, 109323. [CrossRef]
12. Lu, Z.; Lee, P.R.; Yang, H. Kappa-carrageenan improves the gelation and structures of soy protein isolate through the formation of hydrogen bonding and electrostatic interactions. *Food Hydrocoll.* **2023**, *140*, 108585. [CrossRef]
13. Monteiro, S.R.; Lopes-da-Silva, J.A. Effect of the molecular weight of a neutral polysaccharide on soy protein gelation. *Food Res. Int.* **2017**, *102*, 14–24. [CrossRef] [PubMed]
14. Le, X.T.; Rioux, L.E.; Turgeon, S.L. Formation and functional properties of protein–polysaccharide electrostatic hydrogels in comparison to protein or polysaccharide hydrogels. *Adv. Colloid. Interface Sci.* **2017**, *239*, 127–135. [CrossRef] [PubMed]
15. Zhao, H.; Chen, J.; Hemar, Y.; Cui, B. Improvement of the rheological and textural properties of calcium sulfate-induced soy protein isolate gels by the incorporation of different polysaccharides. *Food Chem.* **2020**, *310*, 125983. [CrossRef] [PubMed]
16. Brito-Oliveira, T.C.; Cazado, C.P.; Cavini, A.C.M.; Santos, L.M.; Moraes, I.C.; Pinho, S.C. Cold-set NaCl-induced gels of soy protein isolate and locust bean gum: How the ageing process affect their microstructure and the stability of incorporated beta-carotene. *LWT* **2022**, *154*, 112677. [CrossRef]
17. Chen, N.; Chassenieux, C.; Nicolai, T. Kinetics of NaCl induced gelation of soy protein aggregates: Effects of temperature, aggregate size, and protein concentration. *Food Hydrocoll.* **2018**, *77*, 66–74. [CrossRef]
18. Yu, J.; Wang, Y.; Li, D.; Wang, L.J. Freeze-thaw stability and rheological properties of soy protein isolate emulsion gels induced by NaCl. *Food Hydrocoll.* **2022**, *123*, 107113. [CrossRef]
19. Messerli, F.H.; Hofstetter, L.; Syrogiannouli, L.; Rexhaj, E.; Siontis, G.C.; Seiler, C.; Bangalore, S. Sodium intake, life expectancy, and all-cause mortality. *Eur. Heart J.* **2021**, *42*, 2103–2112. [CrossRef]
20. Cormick, G.; Betrán, A.P.; Metz, F.; Palacios, C.; Beltrán-Velazquez, F.; García-Casal, M.D.L.N.; Peña-Rosas, J.P.; Hofmeyr, G.J.; Belizán, J.M. Regulatory and policy-related aspects of calcium fortification of foods. Implications for implementing national strategies of calcium fortification. *Nutrients* **2020**, *12*, 1022. [CrossRef]
21. Palacios, C.; Hofmeyr, G.J.; Cormick, G.; Garcia-Casal, M.N.; Peña-Rosas, J.P.; Betrán, A.P. Current calcium fortification experiences: A review. *Ann. N. Y. Acad. Sci.* **2021**, *1484*, 55–73. [CrossRef] [PubMed]
22. Teece, L.J.; Faers, M.A.; Bartlett, P. Ageing and collapse in gels with long-range attractions. *Soft Matter* **2011**, *7*, 1341–1351. [CrossRef]
23. Renard, D.; van de Velde, F.; Visschers, R.W. The gap between food gel structure, texture and perception. *Food Hydrocoll.* **2006**, *20*, 423–431. [CrossRef]
24. Buscall, R.; Choudhury, T.H.; Faers, M.A.; Goodwin, J.W.; Luckham, P.A.; Partridge, S.J. Towards rationalising collapse times for the delayed sedimentation of weakly-aggregated colloidal gels. *Soft Matter* **2009**, *5*, 1345–1349. [CrossRef]
25. Bartlett, P.; Teece, L.J.; Faers, M.A. Sudden collapse of a colloidal gel. *Phys. Rev. E* **2012**, *85*, 021404. [CrossRef] [PubMed]
26. Chang, W.Z.; Leong, Y.K. Ageing and collapse of bentonite gels—Effects of Li, Na, K and Cs ions. *Rheol. Acta* **2014**, *53*, 109–122. [CrossRef]
27. Brito-Oliveira, T.C. Effect of the Incorporation of Galactomannans and Solid Lipid Particles Stabilized with Different Surfactants in the Properties and Stability of Cold-Set Gels. Ph.D. Thesis, Faculty of Animal Science and Food Engineering, University of Sao Paulo, Pirassununga, Brazil, 2021.
28. Brito-Oliveira, T.C.; Bispo, M.; Moraes, I.C.; Campanella, O.H.; Pinho, S.C. Stability of curcumin encapsulated in solid lipid microparticles incorporated in cold-set emulsion filled gels of soy protein isolate and xanthan gum. *Food Res. Int.* **2017**, *102*, 759–767. [CrossRef] [PubMed]
29. Brito-Oliveira, T.C.; Bispo, M.; Moraes, I.C.; Campanella, O.H.; Pinho, S.C. Cold-set gelation of commercial soy protein isolate: Effects of the incorporation of locust bean gum and solid lipid microparticles on the properties of gels. *Food Biophys.* **2018**, *13*, 226–239. [CrossRef]
30. Bryant, C.M.; McClements, D.J. Molecular basis of protein functionality with special consideration of cold-set gels derived from heat-denatured whey. *Trends Food Sci. Technol.* **1998**, *9*, 143–151. [CrossRef]
31. Kuhn, K.R.; Cavallieri, Â.L.F.; Da Cunha, R.L. Cold-set whey protein gels induced by calcium or sodium salt addition. *Int. J. Food Sci. Technol.* **2010**, *45*, 348–357. [CrossRef]
32. Vilela, J.A.P.; Cavallieri, Â.L.F.; Da Cunha, R.L. The influence of gelation rate on the physical properties/structure of salt-induced gels of soy protein isolate–gellan gum. *Food Hydrocoll.* **2011**, *25*, 1710–1718. [CrossRef]
33. Marangoni, A.G.; Barbut, S.; McGauley, S.E.; Marcone, M.; Narine, S.S. On the structure of particulate gels—The case of salt-induced cold gelation of heat-denatured whey protein isolate. *Food Hydrocoll.* **2000**, *14*, 61–74. [CrossRef]
34. Guo, J.; Yang, X.Q.; He, X.T.; Wu, N.N.; Wang, J.M.; Gu, W.; Zhang, Y.Y. Limited aggregation behavior of β-conglycinin and its terminating effect on glycinin aggregation during heating at pH 7.0. *J. Agric. Food Chem.* **2012**, *60*, 3782–3791. [CrossRef] [PubMed]

35. Nicolai, T.; Durand, D. Controlled food protein aggregation for new functionality. *Curr. Opin. Colloid. Interface Sci.* **2013**, *18*, 249–256. [CrossRef]
36. Bryant, C.M.; McClements, D.J. Influence of NaCl and $CaCl_2$ on cold-set gelation of heat-denatured whey protein. *J. Food Sci.* **2000**, *65*, 801–804. [CrossRef]
37. Cavallieri, A.L.F.; Da Cunha, R.L. The effects of acidification rate, pH and ageing time on the acidic cold set gelation of whey proteins. *Food Hydrocoll.* **2008**, *22*, 439–448. [CrossRef]
38. Alting, A.C.; Hamer, R.J.; de Kruif, C.G.; Visschers, R.W. Cold-set globular protein gels: Interactions, structure and rheology as a function of protein concentration. *J. Agric. Food Chem.* **2003**, *51*, 3150–3156. [CrossRef] [PubMed]
39. Kuhn, K.R.; Cavallieri, Â.L.F.; Da Cunha, R.L. Cold-set whey protein–flaxseed gum gels induced by mono or divalent salt addition. *Food Hydrocoll.* **2011**, *25*, 1302–1310. [CrossRef]
40. Monteiro, S.R.; Rebelo, S.; Silva, O.A.D.C.; Lopes-da-Silva, J.A. The influence of galactomannans with different amount of galactose side chains on the gelation of soy proteins at neutral pH. *Food Hydrocoll.* **2013**, *33*, 349–360. [CrossRef]
41. de Jong, S.; van de Velde, F. Charge density of polysaccharide controls microstructure and large deformation properties of mixed gels. *Food Hydrocoll.* **2007**, *21*, 1172–1187. [CrossRef]
42. Picone, C.S.; Takeuchi, K.P.; Cunha, R.L. Heat-induced whey protein gels: Effects of pH and the addition of sodium caseinate. *Food Biophys.* **2011**, *6*, 77–83. [CrossRef]
43. Beuschel, B.C.; Culbertson, J.D.; Partridge, J.A.; Smith, D.M. Gelation and emulsification properties of partially insolubilized whey protein concentrates. *J. Food Sci.* **1992**, *57*, 605–609. [CrossRef]
44. Oliver, L.; Scholten, E.; van Aken, G.A. Effect of fat hardness on large deformation rheology of emulsion-filled gels. *Food Hydrocoll.* **2015**, *43*, 299–310. [CrossRef]
45. Chang, Y.Y.; Li, D.; Wang, L.J.; Bi, C.H.; Adhikari, B. Effect of gums on the rheological characteristics and microstructure of acid-induced SPI-gum mixed gels. *Carbohydr. Polym.* **2014**, *108*, 183–191. [CrossRef] [PubMed]

gels

MDPI

Article

Characterization of Beeswax and Rice Bran Wax Oleogels Based on Different Types of Vegetable Oils and Their Impact on Wheat Flour Dough Technological Behavior during Bun Making

Sorina Ropciuc, Florina Dranca, Mircea Adrian Oroian, Ana Leahu, Ancuţa Elena Prisacaru, Mariana Spinei and Georgiana Gabriela Codină *

Faculty of Food Engineering, Stefan cel Mare University of Suceava, 720229 Suceava, Romania;
sorina.ropciuc@fia.usv.ro (S.R.); florina.dranca@usm.ro (F.D.); m.oroian@fia.usv.ro (M.A.O.);
analeahu@fia.usv.ro (A.L.); ancuta.prisacaru@fia.usv.ro (A.E.P.)
* Correspondence: codina@fia.usv.ro

Abstract: Five varieties of vegetable oil underwent oleogelation with two types of wax as follows: beeswax (BW) and rice bran wax (RW). The oleogels were analyzed for their physicochemical, thermal, and textural characteristics. The oleogels were used in the bun dough recipe at a percentage level of 5%, and the textural and rheological properties of the oleogel doughs were analyzed using dynamic and empirical rheology devices such as the Haake rheometer, the Rheofermentometer, and Mixolab. The thermal properties of beeswax oleogels showed a melting peak at a lower temperature for all the oils used compared with that of the oleogels containing rice bran wax. Texturally, for both waxes, as the percentage of wax increased, the firmness of the oleogels increased proportionally, which indicates better technological characteristics for the food industry. The effect of the addition of oleogels on the viscoelastic properties of the dough was measured as a function of temperature. All dough samples showed higher values for G′ (storage modulus) than those of G″ (loss modulus) in the temperature range of 20–90 °C, suggesting a solid, elastic-like behavior of all dough samples with the addition of oleogels. The influence of the beeswax and rice bran oleogels based on different types of vegetable oils on the thermo-mechanical properties of wheat flour dough indicated that the addition of oleogels in dough recipes generally led to higher dough stability and lower values for the dough development time and those related to the dough's starch characteristics. Therefore, the addition of oleogels in dough recipes inhibits the starch gelatinization process and increases the shelf life of bakery products.

Keywords: oleogelation; physicochemical; rheological properties; DSC; rheofermentometer; Mixolab

1. Introduction

Food technologies use solid or semi-solid vegetable oils (shortenings and margarines) in different recipes to obtain baked goods that are soft and have high oxidative stability, an extended shelf life, and good sensory properties. However, their use in bakery product recipes is not desirable from a nutritional point of view due to its content of trans and saturated fatty acids, which may increase the risk of atherosclerosis, blood cholesterol, and other pathologies such as obesity, cardiovascular disease, cancer, and diabetes [1]. Their replacement with other types of fats is difficult to carry out in the food industry due to their technological characteristics, especially their solid properties at room temperature [2]. Nowadays, food producers and scientists are trying to obtain alternative fats without saturated and trans fatty acids that may be used in bakery products [3]. For this purpose, it is necessary to modify the liquid structure of oils into a solid one without changing their nutritional profile. Oleogelation may be a new alternative technology for obtaining semi-solid vegetable oils, and it involves structuring liquid fats using gelling agents; this technology has the potential to provide low saturated fat products with the solid structure that is required in the food industry [1–3]. Different types of oils may be used for oleogel production with a high level of biological activity of the lipids and a high content of essential fatty acids. Olive oil, grape seed oil, hemp seed oil, and walnut oil present a high nutritional value [4], but these oils cannot provide good technological characteristics to foods in their native state. Regarding the chemical composition of the oils, it is mainly composed of triglycerides (97–99%), which are mainly responsible for its biological properties and sensory attributes. Olive oil (OO) has a high content of MUFAs (65–83%), especially oleic acid and polyunsaturated fatty acids (PUFAs), such as linoleic acid, stearic acid, methyl octadecenoate C18:1(trans-9) +methyl oleate C18:1(cis-9), methyl linolelaidate, and methyl docosadienoic C23:0 (cis-16). Hemp seed oil (HO) contains linoleic acid (18:2 n-6) and α-linolenic acid (18:3 n-3) as its major omega-6 and omega-3 polyunsaturated fatty acids (PUFAs), respectively. Linolenic and linoleic fatty acids are known as essential fatty acids because humans cannot produce them themselves and must obtain them from their diet. Grape seed oil (GO) has a high content of PUFAs in the range of 85–90%. Oleic acid and monounsaturated acids (MUFAs) are also highly found in grape seed oil, and saturated fatty acids (SFAs) are present in lower quantities. The main fatty acids in grape seeds have linoleic acid C18:2 cis (n-6)-74.7% and C18:1 cis (n-9)-14.3% oleic acid. Some fatty acid contents of sunflower oil (SO) are palmitic acid 5.0–7.6%, stearic acid 2.7–6.5%, oleic acid 14.0–39.4%, linoleic acid 48.3–74.0%, and linolenic acid- 0–0.3% [5]. The fatty acid composition of walnut oils is as follows: palmitic 6.02–6.4%, palmitoleic 0.11–0.12%, stearic 2.65–2.92%, oleic 15.82–17.12%, linoleic 63.52–62.69%, linolenic 9.66–10.77%, arachidic (C20:0) 0.10–0.11%, and eicosanoid (C20:1) 0.16–0.19% [6]. For this, we propose that different oleogelators may be used for oleogel production with these types of oils such as phospholipids, sorbitan esters, fatty alcohols, mono- and diacylglycerols, ethylcellulose, phytosterols, and plant waxes, e.g., [7]. From these oleogelators, the use of plant waxes is of high interest due to their gel strength, oil binding capacity, abundance, and low cost. Natural waxes derived from seed coatings and plant cuticles as well as those secreted by insects are a mixture of hydrophobic compounds [8]. Different factors such as natural wax type, oil type, and wax concentration level may affect the characteristics of oleogels. For beeswax, the critical concentration of oleogel formation for most vegetable oils is around 6.0% w/w [9], whereas for rice bran it is 5.0% w/w [10,11]. An increase in wax levels decreases the gelation time. At the same time, the fatty acid composition of oils affects oleogel properties. For example, an oleogelator's gelling ability may increase if the vegetable oil used for oleogel production presents a higher amount of saturated fatty acids in its composition [9,12]. According to our previous study, oleogels formulated with olive oil, grape seed oil, hemp seed oil, walnut oil, and sunflower oil using white beeswax (3%, 7%, and 11%) show high potential for saturated fat reduction in fat-based food products [3]. Oleogels have been developed in recent years, and the influence of replacing shortenings with them on the rheological and textural characteristics in various dough matrices, such as those of cookies, sponge cakes, shortcrust biscuits, sweet bread, or croissants, has been studied. According to different studies, the oleogels of candelilla wax with canola oil [7,10] and with rice bran oil reduced dough viscoelasticity [13–15]. Also, the same effect on dough viscoelasticity was reported for oleogels of sunflower oil with candelilla wax and beeswax wax; however, no negative effect on the final quality of bakery products was reported [8,16]. The hardness values for the dough samples increased when different oleogel types such as sunflower oil with carnauba and candelilla wax and canola oil with candelilla wax were incorporated in wheat flour [17]. The viscosity of short doughs decreased when the oleogels of sunflower oil and natural waxes (rice bran wax, beeswax, and candelilla wax) were added to their recipes [16,18]. To our knowledge, there are a few studies on the effects of oleogels on the textural and rheological characteristics of yeast-leavened food systems such as that of breads [2]. Moreover, dough's rheological behavior during the bakery making process is a very complex one that implies different analyses during mixing, fermentation, and baking stages. Previous studies undertaken on the effect of the addition of oleogels into wheat flour on dough's rheological properties are limited since they are focused only on some stages of bakery products process. Our study is a complete one carrying out a full analysis of dough's rheological properties during the mixing, pasting, and fermentation process, which may anticipate the effect of oleogels on bakery's technological process. Also, the aim of this study was to evaluate the effect of replacing shortenings with many types of oleogels such as wax/olive oil, grape seed oil, hemp seed oil, and walnut oil on the textural and rheological characteristics of yeast-leavened dough. These oleogels are diverse ones that may differently affect dough's rheological characteristics. At the same time, the analysis of the oleogel properties that has been carried out during this study may be useful for understanding their behavior during the bakery technological process. This research is a complex and describes the preparation of oleogels from five different assortments of oil structured with beeswax and rice bran wax. The oleogels were described by analyzing the rheological and textural properties. Furthermore, the oleogels were incorporated according to a percentage level of 5% in the dough recipes of bakery products, and the dough's rheological and textural behavior in these products was investigated.

2. Results and Discussion

Liquid oil was successfully structured into solid-like oleogels. The images of the oleogels obtained are shown in Figure 1. The rheological properties of oleogels that were influenced by the frequency and temperature, the melting and freezing behavior of the oleogels, and the textural properties were investigated.

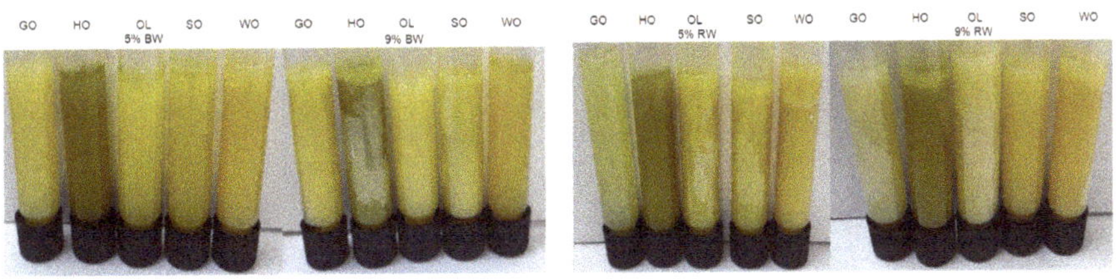

Figure 1. Visual appearance of vegetable oil oleogels with natural waxes.

2.1. *Rheological Characterization of Oleogels according to Frequency*

The frequency-dependent rheological properties were analyzed in order to describe the properties of oleogels. Rheology is one of the most powerful techniques capable of classifying a gelled system based on phase transitions and also of distinguishing between a true gel and a weak gel [19]. The classification of gels by oscillatory rheological measurements can be carried out for strong gels ($G''/G' \leq 0.1$) and not-so-strong gels ($0.1 < G''/G' < 1$), which present less of a gel character [20,21]. Oleogels are systems that have a gel structure, and the condition that they must fulfill is that of the modulus of elasticity being higher than the modulus of viscosity. By comparison, according to Figure 2A,B, beeswax (BW) oleogels had better elastic properties compared with those of rice bran wax (RW) oleogels. At low applied frequency, the elastic component (G') is always higher than the viscous component (G'') for gels and reaches a plateau in the linear viscoelastic region (LVR) [21,22]. With the increasing shear, the deformation of the gel structure takes place, and linearity is only kept in the range of 0.1–1 Hz, corresponding to the LVR range; then, the structure breaks. The moduli of the RW gel had a very low frequency dependence (more or less a linear curve) probably due to the fact that that it has a lower oil binding capacity than that of rice bran wax [20,23]. Moreover, the oleogels had higher G' values than G'' ones, which indicate a solid, elastic-like behavior for all the samples. This behavior was in agreement with the results reported by Patel et al. [22] that also obtained lower loss modulus (G'') values than those of the storage moduli (G') for different types of oleogels. However, the behavior of oleogels varies according to the function of the oil source type [24]. In principle, the fatty acid composition of the oil type regulates the oleogel's flow behaviors. As may be seen, oleogels from grape seed oil, olive oil, and hemp oil present higher values for G' and G''.

According to Zhang et al. [25], oils with more double bonds may be loosely packed and more fluid-like. Also, G″ is highly correlated with the polyunsaturated fatty acid amount, which is at a high level in grape seed or hemp seed oil.

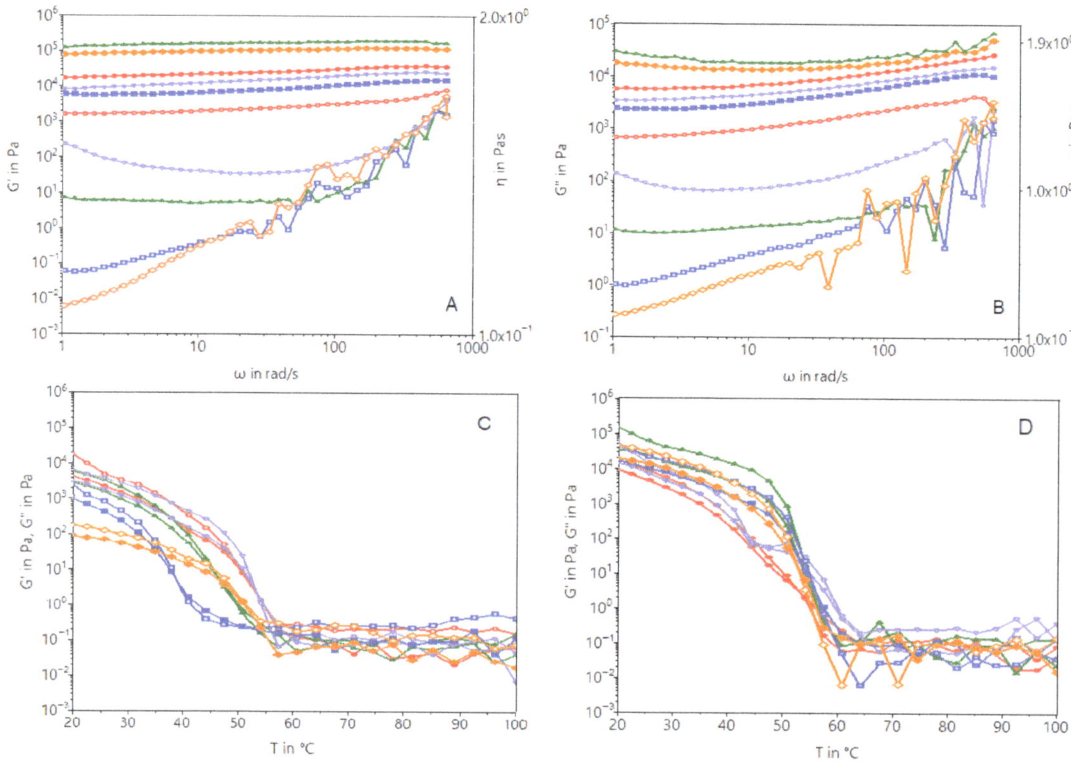

Figure 2. Effect of beeswax and rice bran concentrations on G′ and G″ with the angular frequency (**A**,**B**) and temperature (**C**,**D**). The red line represents the oleogel with grape seed oil (GO), the green line represents the oleogel with hemp seed oil (HO), the purple line represents the oleogel with olive oil (OL), the blue line represents the oleogel with sunflower oil (SO), and the orange line represents the oleogel with walnut oil (WO). Empty symbols represent the addition of 5% beeswax (BW) and solid symbols represent rice bran wax (RW). (**A**)—G′ angular frequency sweeps of oleogels with 5% wax; (**B**)—G″ angular frequency sweep of oleogels; (**C**)—oleogel modules G′ and G″ with 5% BW temperature sweep; (**D**)—oleogels modules G′ and G″ with 9% BW temperature sweep.

The behavior of oleogels at the melting temperature and the transition from the solid phase to the liquid one were analyzed from temperatures that varied in the range of 20–100 °C. The melting curves were approximately similar, and it may be seen that oleogels have a transition temperature of 40–50 °C (SO-5BW and HO-5BW). Similar results were obtained by Tavernier et al. [20] for sunflower oil and hemp seed oil oleogels with beeswax. They correlate the chemical structure of beeswax, which has long-chain wax esters, with a fatty acid portion between C20 and C24 as well as a fatty alcohol portion between C24 and C28, thereby leading to wax crystal formation at lower temperatures than those observed for other oleogelators [26].

The oleogels obtained with the rice bran wax have a very close transition point from the solid phase to the liquid phase. No difference was observed between samples formulated with different oils. An effective gelator must balance its solubility and insolubility in a solvent to achieve proper gelator–gelator and gelator–solvent interactions. The temperature difference observed for the oleogels with sunflower oil and beeswax as well as for the oleogel of hemp seed oil and beeswax at a percentage of 5%, thereby leading to the explanation that these combinations produce oleogels of poor quality. Fundamentally, a lipid structurant must impart solidity to the structured phase as well as prevent the exudation of the structured oil phase from the material. An alternative lipid structurant should exhibit physical properties (hardness/plasticity at a given temperature, melting point, melting profile) that are similar to those of the fat material it is intended to replace.

2.2. Thermal Properties of Oleogels

For the analysis of the thermal properties of oleogels, changes occurring between 20 °C and 90 °C, 90 °C and −60 °C, as well as −60 °C and 100 °C were considered; the values for the DSC parameters of oleogel samples, oils, and waxes—first heating and cooling profiles are presented in Table 1. The thermograms are showed in Figure 3 for oleogels, according to the function of the oil type that was used, for the five different oils, and they are showed separately for beeswax and rice bran wax; the values for the thermal parameters (onset, midpoint, end set, enthalpy, and heat capacity) are presented in Table S1 (Supplementary Materials).

Table 1. DSC parameters of oleogel samples, oils, and waxes—first heating and cooling profiles.

Sample	First Heating, 20 to 90 °C, Melting				First Cooling, 90 to −60 °C, Phase Transition					First Cooling, 90 to −60 °C, Phase Transition 2			
	Onset, °C	Midpoint, °C	End, °C	Enthalpy, J/g	Onset, °C	Midpoint, °C	End, °C	ΔCp, J/(g. °C)	Enthalpy, J/g	Onset, °C	Midpoint, °C	End, °C	Enthalpy, J/g
GO_5BW	43.85	52.47	64.41	2.29	48.82	45.81	43.56	0.0040	0.53	-	-	-	-
GO_9BW	44.71	58.39	68.95	4.52	52.61	49.28	46.59	0.0070	0.93	-	-	-	-
HO_5BW	44.79	53.82	64.10	1.57	49.37	46.10	42.83	0.0060	0.42	-	-	-	-
HO_9BW	42.87	52.98	64.85	4.44	52.82	48.96	46.93	0.0110	1.37	-	-	-	-
OL_5BW	44.05	52.47	61.56	2.28	47.78	45.15	43.70	0.0010	0.61	−57.97	−49.72	−41.35	24.51
OL_9B	43.74	53.85	64.76	4.91	50.96	47.98	46.34	0.0030	1.20	−57.63	−48.89	−39.81	28.76
SO_5BW	44.39	58.72	68.86	2.84	49.71	46.63	43.38	0.0010	0.35	-	-	-	-
SO_9BW	43.42	55.69	67.86	5.17	53.27	49.70	47.45	0.0020	1.22	-	-	-	-
WO_5BW	43.97	52.84	64.53	2.41	49.16	45.49	43.27	0.0050	0.78	-	-	-	-
WO_9BW	43.07	54.27	65.39	5.47	52.29	49.41	47.56	0.0060	0.87	-	-	-	-
GO_5RW	59.91	72.09	81.05	5.48	67.14	63.36	59.55	0.0110	1.00	-	-	-	-
GO_9RW	58.25	72.31	82.02	7.78	70.45	64.76	63.21	0.0120	5.56	-	-	-	-
HO_5RW	58.81	71.53	80.61	5.92	70.11	65.32	62.26	0.0160	2.61	-	-	-	-
HO_9RW	54.19	72.56	81.95	14.09	75.17	68.75	67.20	0.0270	6.36	-	-	-	-
OL_5RW	45.12	70.57	84.19	9.67	68.00	63.74	61.39	0.0080	1.71	−58.51	−50.41	−41.65	24.38
OL_9RW	44.04	70.50	84.56	14.77	69.51	62.66	61.38	0.0350	6.97	−58.43	−50.20	−42.45	27.82
SO_5RW	43.59	70.13	84.06	9.43	68.01	62.92	59.86	0.0020	1.38	-	-	-	-
SO_9RW	43.34	70.42	83.27	17.82	70.92	65.70	62.79	0.0150	4.34	-	-	-	-
WO_5RW	43.59	67.36	80.16	4.45	57.56	54.65	52.46	0.0001	0.53	-	-	-	-
WO_9RW	42.99	71.01	82.81	21.27	73.97	66.72	65.07	0.0160	11.13	-	-	-	-
GO	-	-	-	-	−16.43	−14.68	−12.59	0.0010	0.19	−53.28	−36.01	−31.80	7.39
HO	-	-	-	-	-	-	-	-	-	-	-	-	-
OL	-	-	-	-	−11.09	−13.88	−15.42	0.0020	1.23	−40.68	−48.84	−58.54	29.97
SO	-	-	-	-	-	-	-	-	-	-	-	-	-
WO	-	-	-	-	-	-	-	-	-	-	-	-	-
BW	34.68	67.28	83.91	185.05	67.53	60.49	59.34	0.0230	38.77	-	-	-	-
RW	77.12	82.00	84.44	20.71	81.96	76.27	75.47	0.0030	55.12	61.12	57.69	47.29	20.44

GO_5BW, GO_9BW, GO_5RW, and GO_9RW—oleogel with grape seed oil and 5% and 9% beeswax and, respectively, rice bran wax; HO_5BW, HO_9BW, HO_5RW, and HO_9RW—oleogel with hemp seed oil and 5% and 9% beeswax and, respectively, rice bran wax; OL-5BW, OL-9BW, OL-5RW, and OL_9RW—oleogel with olive oil and 5% and 9% beeswax and, respectively rice bran wax; SO_5BW, SO_9BW, SO_5RW, SO_9RW—oleogel with sunflower oil and 5% and 9% beeswax and, respectively, rice bran wax; WO_5BW, WO_9BW, WO_5RW, and WO_9RW—oleogel with walnut oil and 5% and 9% beeswax and, respectively, rice bran wax.

As can be observed in Figure 3 and in Figure S1 in the Supplementary Materials, the first heating profile (from 20 °C to 90 °C) of the oleogels and waxes showed similar thermal behaviors and contained one melting peak. As this peak was not found in the thermograms of the oil samples, any changes during this heating profile were attributed to the oleogelator (beeswax or rice bran wax). The melting peak was recorded at a lower temperature for oleogels with beeswax in comparison with oleogels containing rice bran wax, respectively. This was due to the lower melting temperature of beeswax when compared with that of rice bran wax, which was in accordance with the results of previous studies [11,27]. For all the oleogel samples, the melting peak was observed at a lower temperature than that of the melting peak of the wax used, and the melting enthalpy was lower when the proportion of wax in oleogels was 5%. The lower peak temperature and enthalpy at a lower oleogelator concentration were due to the dilution effect of the solvent, namely the oil used, which was also reported for oleogels with corn oil and rice bran oil [27,28].

The first cooling profile comprised two exothermic peaks, one corresponding to the phase transition, which was recorded for all oleogel samples and waxes, and one attributed to crystallization, which was only characteristic of oleogels containing olive oil and of the olive oil sample. Both BW and the oleogels with BW had a lower phase transition enthalpy in comparison with that of the RW oleogels and rice bran wax, indicating the lower crystallinity of this oleogelator, with these data being in agreement with previous studies [29]. The second exothermic peak that had a higher enthalpy was linked to the phase transition of the low melting of the highly unsaturated oil fraction of olive oil (oleic, linoleic, and palmitic acids) [30].

During the second heating and cooling profile, several endothermic and exothermic peaks were observed for the thermograms of the oleogel samples. Firstly, all the oleogels had a melting peak in the range between −40 and 0 °C, which was consistent with the endothermic peak that was recorded for the oil samples; this endothermic peak was also reported by other authors that studied the thermal behavior of vegetable oils [31,32] and was mostly related to the high content of low-melting triacylglycerols and the high proportion of mono- and polyunsaturated fatty acids that are the main components of the vegetable oils used to prepare oleogels [33]. The second melting peak was similar to the peak recorded in the first heating profile; however, the melting enthalpy was lower when compared with the first heating, thereby suggesting that after the first heating and cooling profile, the structure of the oleogel changed. In the second cooling phase, the same endothermic peak corresponding to the phase transition was found in all oleogel samples, and the enthalpy and heat capacity values were overall higher than those determined for the first cooling profile

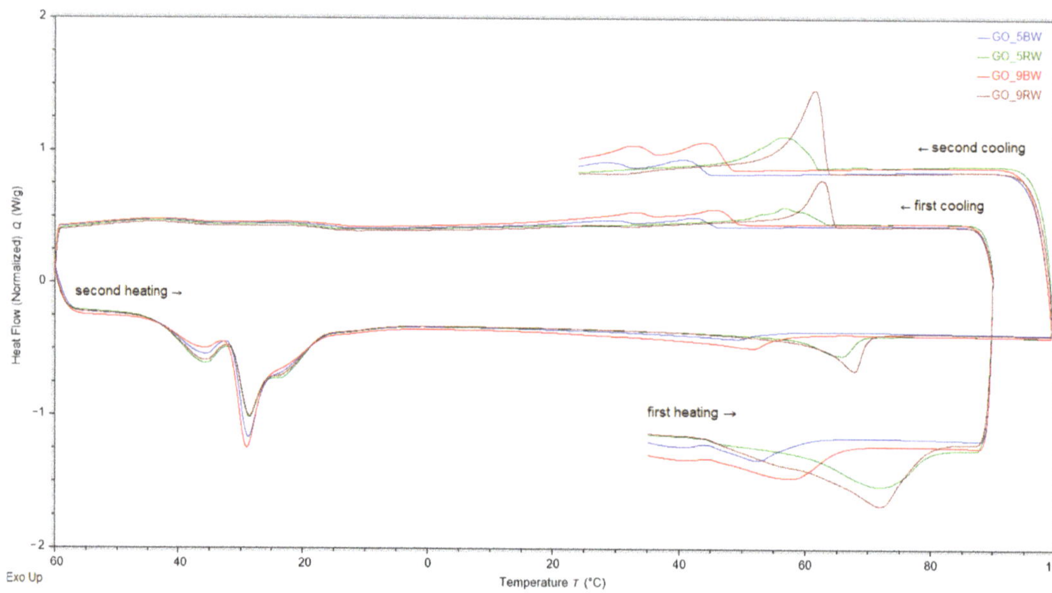

(A)

Figure 3. *Cont.*

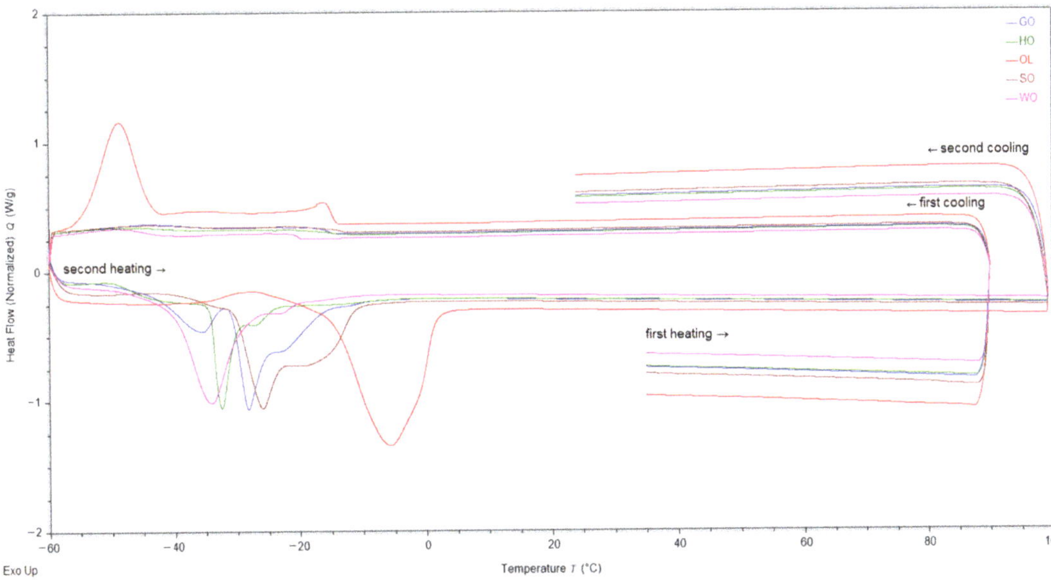

(**B**)

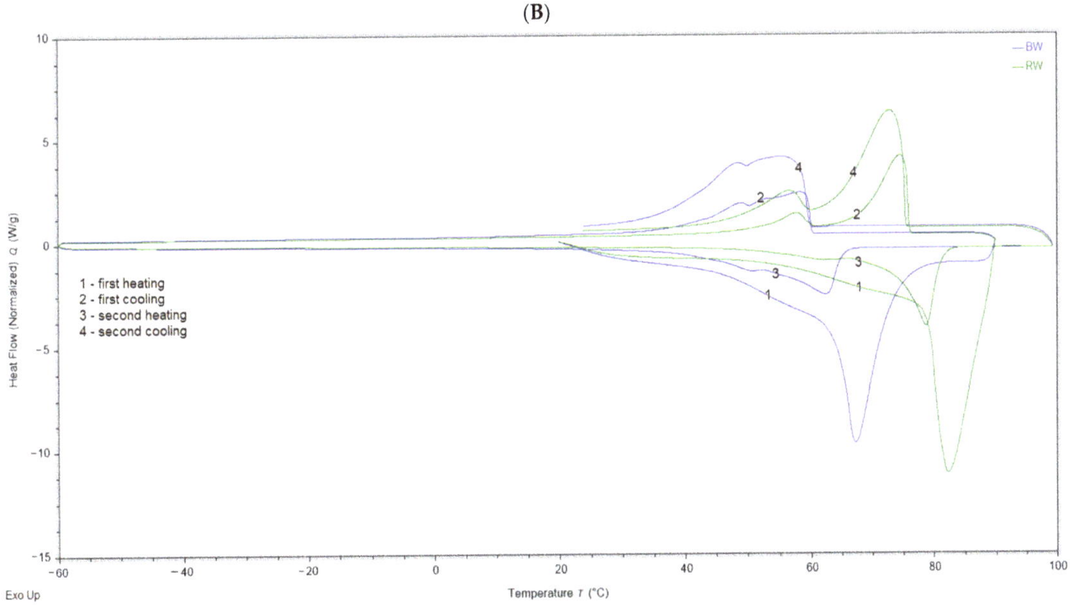

(**C**)

Figure 3. Thermograms of oleogels, oils, and waxes for heating (→) and cooling (←) profiles: (**A**) oleogels with grape oil, (**B**) oils, and (**C**) beeswax and rice bran wax.

2.3. Thermogravimetric Analysis of Oleogels

The thermal degradation of oleogels based on the oil type (grape seed, hemp, olive, sunflower, and walnut oil) and also the wax type (beeswax and rice bran wax) that was used for the formulation of them was studied by thermogravimetric analysis (TGA). Figure S2 in the Supplementary Materials presents the TGA curves, in the form of weight loss versus sample temperature, for the analyzed oleogels. Table 2 shows the values for the onset temperature (T_{onset}) and the endset temperature (T_{endset}) of the induced mass loss, the maximum weight loss rate (T_{max}), the mass loss at the end of each decomposition (ΔW), and the residue that remains at 600 °C, which were calculated from thermograms using the method described by Sánchez et al. [34]. Excluding the initial mass loss at 150–200 °C, which is around 5%, as a result of water evaporation, another essential range is that oleogel decomposition took place in two well-defined stages. Therefore, the first stage (also known as the main stage) occurred between 295 and 480 °C due to the decomposition of the part of the oleogel molecules that depends on the percentage of saturated and unsaturated fatty acids present in the mass of total fat [35]. Moreover, this stage is also associated with the thermal volatilization of the vegetable oil used for the formulation of the oleogels [36]. The sample containing olive oil and 5% beeswax displays a slightly lower thermal stability (lower T_{onset}, 297.26 °C), and the maximum degradation rate performed at a lower temperature (lower T_{endest}, 451.10 °C). The second stage (between 480 and 555 °C) may be attributed to the degradation of the long-chain molecules (e.g., long-chain fatty acids, triacylglycerols) present in wax, increasing the possibility of crystallization [37]. The sample formulated with walnut oil and rice bran wax had the highest residue remaining at 600 °C (26.13%) in comparison with the beeswax sample; this can be explained by the fact that rice bran wax has a higher melting point compared with that of beeswax (this fact is supported by DSC analysis). The thermogravimetric data also show that the type of vegetable oil has a significant influence on the residue remaining at 600 °C, with lower values for olive oil and sunflower oil. Therefore, it can be observed that the incorporation of 9% beeswax and rice bran wax in the oleogel formulation has led to an improvement in thermal stability.

Table 2. TGA parameters of oleogel samples.

Sample	T_{onset} (°C)	T_{endset} (°C)	T_{final} (°C)	ΔW (%)	Residue at 600 °C (%)
GO-5BW	323.11	473.48	484.76	83.89	16.11
GO-9BW	344.37	456.29	489.10	91.72	8.28
HO-5BW	355.90	464.09	494.47	86.56	13.44
HO-9BW	351.20	462.46	492.26	83.79	16.21
OL-5BW	297.26	451.10	497.00	93.96	6.04
OL-9BW	325.63	453.64	484.99	95.91	4.09
SO-5BW	365.27	467.64	499.52	96.37	3.63
SO-9BW	352.34	470.41	506.16	94.79	5.21
WO-5BW	347.49	478.81	519.74	93.00	7.00
WO-9BW	379.07	472.20	503.63	91.53	8.47
GO-5RW	355.46	470.95	503.00	92.55	7.45
GO-9RW	377.46	465.07	493.52	97.06	2.94
HO-5RW	378.64	471.00	498.89	89.61	10.39
HO-9RW	381.36	473.39	500.16	89.06	10.94
OL-5RW	360.34	463.31	555.45	97.09	2.91
OL-9RW	349.41	461.57	502.37	97.38	2.62
SO-5RW	351.67	467.95	503.95	97.66	2.34
SO-9RW	354.38	467.06	499.21	97.80	2.20
WO-5RW	380.29	472.22	513.74	73.87	26.13
WO-9RW	379.84	476.15	504.58	93.05	6.95

GO-5BW, GO-9BW GO-5RW, and GO-9RW—oleogel with grape seed oil and 5% and 9% beeswax and, respectively, rice bran wax; HO-5BW, HO-9BW, HO-5RW, and HO-9RW—oleogel with hemp seed oil and 5% and 9% beeswax and, respectively, rice bran wax; OL-5BW, OL-9BW, OL-5RW, and OL-9RW—oleogel with olive oil and 5% and 9% beeswax and, respectively, rice bran wax; SO-5BW, SO-9BW, SO-5RW, and SO-9RW—oleogel with sunflower oil and 5% and 9% beeswax and, respectively, rice bran wax; WO-5BW, WO-9BW, WO-5RW, and WO-9RW—oleogel with walnut oil and 5% and 9% beeswax and, respectively, rice bran wax.

2.4. Texture Profile Analysis of Oleogels

The textural properties of oleogels (Table 3) are as important as the thermal properties and define the phenomena of gel stability and oil retention in the dough structure as well as their binding capacity. If the oleogels are of high hardness at a temperature of 20–22 °C, the possibility of incorporation into the dough is hindered. There is a risk that the structure of the oleogel does not allow for the incorporation of the oleogel during mixing and that it remains in the dough as particles. Of course, this can be prevented by bringing the oleogels to a temperature that ensures their complete incorporation into the dough mass. For both waxes, as the percentage of wax increased, the firmness of the oleogels increased proportionally ($p < 0.001$). Different values for the textural parameters were observed between the oil varieties used for oil freezing. For example, sunflower oil (SO), walnut oil (WO), and hemp seed oil with rice bran wax (RW) lead to oleogels with lower hardness (24–26 g) and higher stickiness (9.9–8.8%) compared with the samples obtained with grapeseed oil (GO) and olive oil (OL). The adhesiveness of the oleogel samples was close in value, and oleogels with higher stickiness were obtained in walnut oil and beeswax oleogels (WO-5BW) and also in sunflower oil and rice bran wax (RW) oleogels. The addition of 9% wax resulted in oleogels with higher hardness and lower stickiness. For most oleogels, it can be seen that the two types of wax form similar oleogels with different characteristics.

Table 3. TPA profile analysis of oleogels.

Sample	Hardness (g)	Springiness (%)	Stickiness (%)	Adhesiveness (g.s)	Cohesiveness (%)
GO-5BW	30 ^{de}	0.9 ^{ab}	−8.6 ^{b}	−2.3 ^{a}	0.63 c
HO-5BW	26 ^{fg}	0.9 ^{ab}	−9.81 ^{b}	−3.8 ^{a}	0.56 ^{c}
OL-5BW	32 ^{cd}	1.0 ^{ab}	−7.8 ^{ab}	−3.6 ^{a}	0.55 ^{d}
SO-5BW	30 ^{de}	0.8 ^{b}	−6.9 ^{ab}	−2.8 ^{a}	0.47 ^{ef}
WO-5BW	29 ^{def}	1.2 ^{a}	−9.8 ^{b}	−3.5 ^{a}	0.42 ^{f}
GO-9BW	32 ^{cd}	0.9 ^{ab}	−7.9 ^{ab}	−3.4 ^{a}	0.88 ^{a}
HO-9BW	39 ^{a}	1.1 ^{ab}	−8.6 ^{b}	−2.4 ^{a}	0.82 ^{a}
OL-9BW	37 ^{ab}	1.0 ^{ab}	−8.5 ^{b}	−3.2 ^{a}	0.78 ^{a}
SO-9BW	31 ^{cde}	1.0 ^{ab}	−9.2 ^{b}	−2.6 ^{a}	0.84 ^{a}
WO-9BW	30 ^{cd}	0.8 ^{b}	−9.7 ^{b}	−2.7 ^{a}	0.77 ^{ab}
GO-5RW	36 ^{ab}	1.2 ^{a}	−8.5 ^{b}	−3.1 ^{a}	0.33 ^{g}
HO-5RW	26 ^{fg}	1.0 ^{ab}	−8.6 ^{b}	−3.8 ^{a}	0.52 ^{de}
OL-5RW	28 ^{ef}	1.0 ^{ab}	−9.0 ^{b}	−2.9 ^{a}	0.37 ^{g}
SO-5RW	24 ^{gh}	0.8 ^{b}	−9.8 ^{b}	−3.4 ^{a}	0.41 ^{f}
WO-5RW	26 ^{fg}	0.8 ^{b}	−8.5 ^{b}	−3.6 ^{a}	0.24 ^{h}
GO-9RW	36 ^{ab}	0.9 ^{ab}	−8.6 ^{b}	−3.1 ^{a}	0.67 ^{ab}
HO-9RW	32 ^{cd}	0.9 ^{ab}	−9.9 ^{b}	−2.8 ^{a}	0.42 ^{f}
OL-9RW	34 ^{bc}	1.1 ^{ab}	−8.5 ^{b}	−2.3 ^{a}	0.36 ^{g}
SO-9RW	36 ^{ab}	1.2 ^{a}	−9.1 ^{b}	−2.4 ^{a}	0.22 ^{h}
WO-9RW	30 ^{de}	1.0 ^{ab}	−9.1 ^{b}	−2.9 ^{a}	0.21 ^{h}
F-value	57.62 ***	4.82 ***	2.88 ***	0.58 ***	12.24 ***

GO_5BW, GO_9BW GO_5RW, and GO_9RW—oleogel with grape seed oil and 5% and 9% beeswax and, respectively, rice bran wax; HO_5BW, HO_9BW, HO_5RW, and HO_9RW—oleogel with hemp seed oil and 5% and 9% beeswax and, respectively, rice bran wax; OL_5BW, OL_9BW, OL_5RW, and OL_9RW—oleogel with olive oil and 5% and 9% beeswax and, respectively, rice bran wax; SO_5BW, SO_9BW, SO_5RW, and SO_9RW—oleogel with sunflower oil and 5% and 9% beeswax and, respectively, rice bran wax, WO_5BW, WO_9BW, WO_5RW, and WO_9RW—oleogel with walnut oil and 5% and 9% beeswax and, respectively, rice bran wax. Means in the same column with different letters for each sample type are significantly different ($p < 0.05$), and the F-value represents Tukey's HSD test,*** strong correlation.

2.5. Rheological Characterization of Dough during Heating

The effects of the addition of oleogels at a 5% addition level in wheat flour on the viscoelastic properties of dough were measured as a function of temperature. All dough samples showed higher values for G′ (storage modulus) than those of G″ (loss modulus) in the temperature range of 20–90 °C (Figure 4). In the initial phase, the values for G′ and the G″ modulus decreased as the temperature increased, probably due to the protein denaturation that cannot retain water from the dough system. Gradually, the water release begins to be absorbed by starch granules that gelatinize, increasing the G′ and G″ values.

As may be seen, the G′ values for the dough samples with the addition of oleogels with 5% beeswax are higher than those obtained for the dough samples with oleogels in which the percentage of wax was 9%. Jung et al. [11] formulated mixtures between rice bran oil and candelilla wax oleogels and introduced them into sweet breads, determining the extensibility of the dough accordingly. In the case of extensibility, the dough samples became more extensible with increasing oleogel levels in the dough recipe. In our research, the oleogel was used at a percentage of 5% in the leavened dough, which makes the results differ because of the type of oil used to formulate the oleogel, the wax used for oleogelation, and the percentage of the added wax in the wheat flour. In Figure 4A,C, the doughs with oleogel from sunflower oil (DSO-5BW) showed the highest values for the modulus of elasticity in the case of the dough samples with the oleogel with the beeswax addition. Comparatively, doughs with the oleogel from walnut oil (DWO-9RW) show high elastic behavior in samples formulated with 9% rice bran wax. The gelatinization temperature (60–80 °C) of the starch was influenced by the addition of the oleogel in the dough and not by the percentage of the added wax. The doughs obtained with rice bran wax (RW-9) showed the lowest values for the modules during heating. The values for the G′ G″ modules decreased with the addition of the oleogel with a wax content of 9%. This results is not an desirable one for fermented doughs because a tenacious dough with low elasticity cannot develop a product with a high specific volume and a crumb with fine porosity [38,39].

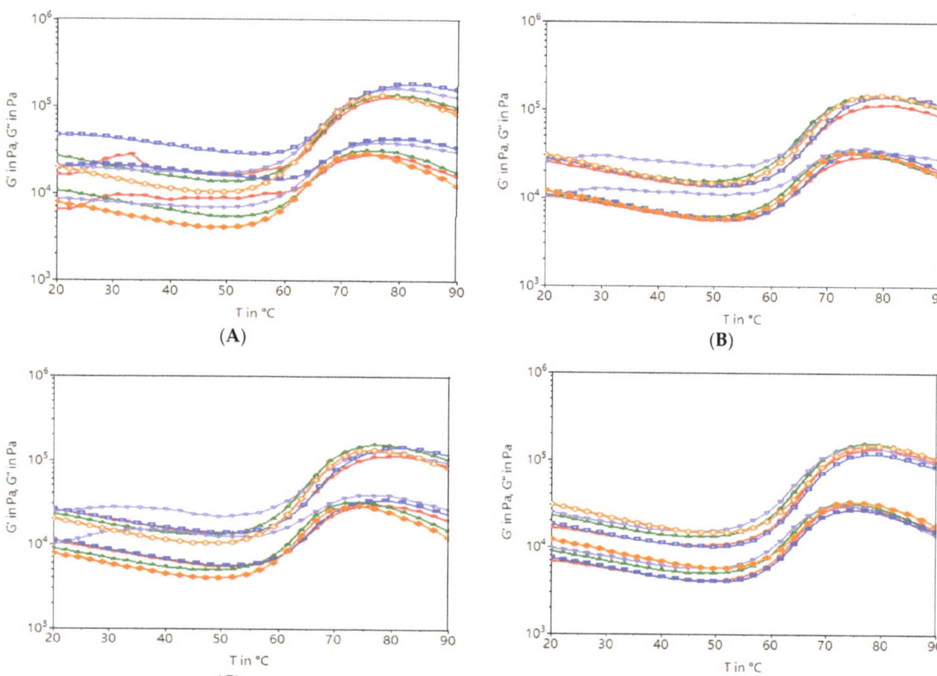

(A) (B) (C) (D)

Figure 4. Graphical representation of the viscoelastic modules for dough samples with the addition of oleogels. (**A**,**B**)—describe the viscoelastic modules for doughs obtained with oleogels with 5% and 9% BW; (**C**,**D**)—represent the viscoelastic modules for doughs with the addition of 5% and 9% RW. The red line represents the dough with oleogel from grape seed oil (GO), the green line represents the dough with oleogel from hemp seed oil (HO), the purple line represents the dough with oleogel from olive oil (OL), the blue line represents the dough with oleogel from sunflower oil (SO), and the orange line represents the dough with oleogel from walnut oil (WO). The empty symbols represent the module G′ and the filled symbols represent G″.

2.6. Rheological Properties of Dough during Fermentation

The rheological properties of wheat flour dough with an oleogel addition at a 5% addition level in wheat flour during fermentation are presented in Table 4. The dough samples with different oil and wax percentages of 5 and 9% showed significant variations in the maximum height of gas production (H'm). The H'm values for the samples with RW were between 42.20 mm and 47.11 mm compared with the dough samples to which oleogels based on beeswax (BW) were added, whereas the maximum height reached the value of 39.6 mm. Regarding the total volume of CO_2 produced during fermentation (VT) and the volume of gas retained in the dough at the end of the test (VR), the highest values were obtained for the dough sample with oleogel obtained from olive oil and rice bran wax at a 9% addition (DOL-9RW), while the lowest ones were obtained for the dough samples with the oleogel obtained from sunflower seed oil and a 5% beeswax addition (DSO-5BW) and the oleogel with hemp seed oil and beeswax rice bran (DHO-5RW). The VT and VR values increased in the dough samples that contained oleogels with olive oil and 9% rice bran wax (RW) and beeswax (BW). This behavior is related to the viscous and elastic responses of the dough samples with the addition of oleogels, which may lead to a lower or a higher gas retention capacity. During dough fermentation, gas expansion increases the surface area and adsorption of lipid crystals. Crystalline lipids in oleogels with higher melting points that remain solid at the end of fermentation are the most effective at trapping gases. However, when too many lipids in solid form are present in the dough, its ability to grow during fermentation is inhibited [40]. Furthermore, the addition of oleogels containing solid lipids in an appropriate amount can strengthen the dough and thereby improve gas retention. The retention coefficient showed higher values for the dough samples obtained with oleogels from grape seed oil, walnut oil, and high wax BW/RW oleogels from olive oil and sunflower oil. It is well known that in bread making, lipids improve gas retention and thus increase its volume. According to Jung et al. [41], oleogels with solid-like properties are very effective in increasing the dough volume, which means it can improve the CR value and also the H'm and VT ones. Taking into account the fact that the CR is the ratio of the VR and VT values, it can be concluded that in some cases where the CR showed lower values compared with those of the control sample, the capacity of the dough to retain gases is lower than the amount of gases formed during the dough fermentation process [42].

Table 4. Rheological properties during fermentation of the dough samples.

Sample	H'm (mm)	VT (mL)	VR (mL)	CR (VR/VT), %
DGO-5BW	36.65 ^{gh}	1250 ^{cd}	1065 ^{g}	88.32 ^{ef}
DHO-5BW	38.36 ^{fgh}	1263 ^{cd}	1096 ^{b}	86.71 ^{f}
DOL-5BW	38,54 ^{ef}	1131 ^{fg}	1025 ^{l}	90.64 ^{cd}
DSO-5BW	38.36 ^{efg}	986 ^{j}	941 ^{q}	95.50 ^{a}
DWO-5BW	34.40 ^{i}	1275 ^{b}	1030 ^{j}	80.80 ^{h}
DGO-9BW	37.84 ^{efgh}	1117 ^{g}	1043 ^{i}	93.30 ^{b}
DHO-9BW	34.49 ^{i}	1133 ^{fg}	1013 ^{n}	89.40 ^{de}
DOL-9BW	33.99 ^{i}	1239 ^{d}	1090 ^{d}	88.05 ^{ef}
DSO-9BW	37.48 ^{h}	1239 ^{d}	1094 ^{c}	88.32 ^{ef}
DWO-9BW	39.60 ^{e}	1212 ^{e}	1079 ^{e}	89.10 ^{de}
DGO-5RW	45.21 ^{a}	1117 ^{g}	1043 ^{i}	93.30 ^{b}
DHO-5RW	42.92 ^{bcd}	905 ^{k}	750 ^{r}	82.93 ^{g}
DOL-5RW	44.51 ^{ab}	1046 ^{i}	985 ^{o}	94.22 ^{ab}
DSO-5RW	42.20 ^{d}	1236 ^{d}	1074 ^{f}	86.94 ^{f}
DWO-5RW	45.42 ^{a}	1086 ^{h}	1016 ^{m}	93.63 ^{b}
DGO-9RW	42.44 ^{cd}	1127 ^{g}	1028 ^{k}	91.22 ^{c}
DHO-9RW	44.23 ^{abc}	1151 ^{f}	1078 ^{e}	90.81 ^{cd}
DOL-9RW	46.00 ^{a}	1364 ^{a}	1220 ^{a}	88.02 ^{ef}
DSO-9RW	42.60 ^{cd}	1209 ^{d}	1062 ^{h}	87.84 ^{ef}

Table 4. *Cont.*

Sample	H'm (mm)	VT (mL)	VR (mL)	CR (VR/VT), %
DWO-9RW	42.62 ^{cd}	1049 ^{i}	982 ^{p}	93.51 ^{b}
F-value	47.114 ***	262.27 ***	23,358.08 ***	42.57 ***

DGO_5BW, DGO_9BW DGO_5RW, and DGO_9RW—dough samples with 5% addition of oleogel with grape seed oil and 5% and 9% beeswax and, respectively, rice bran wax; DHO_5BW, DHO_9BW, DHO_5RW, and DHO_9RW—dough samples with 5% addition of oleogel with hemp seed oil and 5% and 9% beeswax and, respectively, rice bran wax; DOL_5BW, DOL_9BW, DOL_5RW, and DOL_9RW—dough samples with 5% addition of oleogel with olive oil and 5% and 9% beeswax and, respectively, rice bran wax; DSO_5BW, DSO_9BW, DSO_5RW, and DSO_9RW—dough samples with 5% addition of oleogel with sunflower oil and 5% and 9% beeswax and, respectively, rice bran wax; DWO_5BW, DWO_9BW, DWO_5RW, and DWO_9RW—dough samples with 5% addition of oleogel with walnut oil and 5% and 9% beeswax and, respectively, rice bran wax. Means in the same column with different letters for each sample type are significantly different ($p < 0.05$), and F-value represents Tukey's HSD test,*** strong correlation.

2.7. Rheological Characterization of Dough from Wheat Flour and Different Types of Oleogels Using Mixolab

As Table 5 shows, different rheological data were obtained depending on the oleogel type used in the dough recipe. In the initial Mixolab mixing stage, stability (ST), dough development time (DDT), and C1 torque were determined. According to the data obtained, the DDT and ST are significantly different ($p < 0.05$) for the samples with the oleogel and shortening addition compared with those of the dough only obtained from wheat flour (control sample). Generally, compared with the control sample, the oleogel and shortening addition in wheat flour increased dough stability and decreased dough development time values. The sample with the shortening addition in wheat flour presented a similar behavior to that of the dough samples, which have oleogels incorporated in their recipe. Non-significant differences ($p < 0.05$) between the wheat flour dough with the 5% shortening addition (DWF-5SG) and the oleogel with olive oil and 5% beeswax (DOL-5BW) were obtained for the dough rheological properties of the mixing values. The increase in the stability value may be due to the lipid content from the oleogel and shortening samples, which may form lipoprotein complexes between the gluten, starch, and hydrophobic components, thereby leading to a more compact and stable dough [43]. Also, dough with the solid oleogel and shortening addition contain expandable and thin gluten films that favor air incorporation and dough stabilization during mixing [44]. The decrease in the DDT values with the oleogel addition is in agreement with those reported by Jung et al. [41] who concluded that a high amount of an oleogel addition to wheat flour dough leads to less resistance against mixing. The C1 torque value was 1.1 ± 0.04 Nm and represented the optimal dough consistency (C1 = 1.1 Nm) during mixing. Generally, when compared with the control sample, the Mixolab torques C2, C3, C4, and C5 decreased in the dough samples with the oleogel and shortening, and these data were in agreement with those reported by other authors [41,45]. According to them, vegetable oil dispersed all through the dough mixing in the form of oleogels or shortening, thereby leading to a soft dough that is less viscous and decreased the Mixolab torques during heating and cooling. These data are also in agreement with those reported by Oh et al. [46] who concluded that the addition of oleogels obtained using the organogelators of rice bran wax and beeswax reduced the viscosity of the dough samples, with this reduction being observed as higher for oleogels with rice bran wax compared with those obtained from beeswax. This different behavior may be due to the wax composition, which the in case of beeswax comprises 58% wax ester and more than 26% hydrocarbon, whereas in the case of rice bran wax it comprises 90% wax esters [15]. As the hydrogenated level of fat becomes higher, the dough becomes harder.

Table 5. Mixolab parameter values for wheat flour with different types of wax oleogel additions.

Samples	ST (min)	DDT (min)	C1 (N·m)	C2 (N·m)	C3 (N·m)	C4 (N·m)	C5 (N·m)	C12 (N·m)	C32 (N·m)	C54 (N·m)
WF	9.26 ± 0.05 ^{g}	1.73 ± 0.05 ^{i}	1.08 ± 0.03 ^{ab}	0.48 ± 0.01 ^{g}	1.73 ± 0.02 ^{g}	2.20 ± 0.33 ^{m}	4.43 ± 0.01 ^{o}	0.59 ± 0.30 ^{a}	1.24 ± 0.15 ^{f}	2.23 ± 0.04 ^{j}
WF-5SG	9.68 ± 0.10 ^{k}	1.57 ± 0.02 ^{fgh}	1.10 ± 0.01 ^{ab}	0.46 ± 0.01 ^{efg}	1.59 ± 0.01 ^{f}	1.58 ± 0.01 ^{f}	2.84 ± 0.01 ^{d}	0.62 ± 0.02 ^{ab}	1.13 ± 0.02 ^{f}	1.26 ± 0.01 ^{de}
DGO-5BW	9.56 ± 0.05 ^{ijk}	1.12 ± 0.01 ^{b}	1.13 ± 0.15 ^{b}	0.45 ± 0.01 ^{defg}	1.16 ± 0.02 ^{e}	1.11 ± 0.01 ^{d}	2.91 ± 0.01 ^{h}	0.68 ± 0.17 ^{bc}	0.70 ± 0.13 ^{e}	1.79 ± 0.01 ^{f}
DHO-5BW	8.80 ± 0.10 ^{f}	1.59 ± 0.02 ^{gh}	1.07 ± 0.02 ^{ab}	0.44 ± 0.05 ^{cdefg}	1.05 ± 0.01 ^{bcde}	1.59 ± 0.01 ^{g}	2.87 ± 0.00 ^{def}	0.64 ± 0.55 ^{abc}	0.62 ± 0.54 ^{bcde}	1.28 ± 0.01 ^{e}
DOL-5BW	9.66 ± 0.05 ^{jk}	1.58 ± 0.02 ^{fgh}	1.10 ± 0.01 ^{ab}	0.46 ± 0.00 ^{efg}	1.01 ± 0.01 ^{bc}	1.66 ± 0.12 ^{g}	2.81 ± 0.01 ^{c}	0.65 ± 0.02 ^{abc}	0.55 ± 0.01 ^{b}	1.14 ± 0.01 ^{b}
DSO-5BW	9.66 ± 0.05 ^{jk}	1.59 ± 0.02 ^{gh}	1.09 ± 0.01 ^{ab}	0.44 ± 0.02 ^{bcdef}	1.10 ± 0.00 ^{bcde}	1.06 ± 0.00 ^{c}	2.98 ± 0.01 ^{i}	0.65 ± 0.28 ^{abc}	0.66 ± 0.23 ^{bcde}	1.91 ± 0.01 ^{i}
DWO-5BW	9.70 ± 0.10 ^{k}	1.58 ± 0.02 ^{fgh}	1.10 ± 0.01 ^{ab}	0.45 ± 0.00 ^{efg}	1.01 ± 0.00 ^{abc}	1.66 ± 0.00 ^{gh}	2.80 ± 0.01 ^{c}	0.65 ± 0.00 ^{abc}	0.55 ± 0.01 ^{b}	1.14 ± 0.01 ^{b}
DGO-9BW	9.60 ± 0.10 ^{jk}	1.47 ± 0.02 ^{de}	1.09 ± 0.01 ^{ab}	0.42 ± 0.00 ^{bcdef}	1.58 ± 0.00 ^{f}	1.52 ± 0.01 ^{e}	2.72 ± 0.02 ^{b}	0.67 ± 0.00 ^{abc}	1.15 ± 0.00 ^{f}	1.20 ± 0.15 ^{c}
DHO-9BW	8.36 ± 0.05 ^{e}	1.47 ± 0.02 ^{de}	1.08 ± 0.01 ^{ab}	0.41 ± 0.01 ^{bcd}	1.12 ± 0.01 ^{cde}	1.71 ± 0.12 ^{j}	2.90 ± 0.01 ^{gh}	0.68 ± 0.01 ^{bc}	0.71 ± 0.02 ^{e}	1.19 ± 0.01 ^{c}
DOL-9BW	9.76 ± 0.05 ^{k}	0.88 ± 0.02 ^{a}	1.08 ± 0.01 ^{ab}	0.42 ± 0.01 ^{bcdef}	2.22 ± 0.02 ^{h}	2.16 ± 0.01 ^{l}	2.66 ± 0.01 ^{a}	0.66 ± 0.14 ^{abc}	1.80 ± 0.37 ^{g}	0.49 ± 0.01 ^{a}
DSO-9BW	4.30 ± 0.10 ^{c}	1.49 ± 0.02 ^{ef}	1.10 ± 0.01 ^{ab}	0.40 ± 0.01 ^{bc}	1.10 ± 0.01 ^{bcde}	1.05 ± 0.01 ^{c}	2.88 ± 0.01 ^{defg}	0.70 ± 0.01 ^{c}	0.70 ± 0.10 ^{e}	1.82 ± 0.01 ^{fg}
DWO-9BW	7.13 ± 0.05 ^{d}	1.37 ± 0.03 ^{cd}	1.07 ± 0.04 ^{ab}	0.46 ± 0.02 ^{efg}	1.03 ± 0.00 ^{bcd}	0.97 ± 0.02 ^{b}	3.45 ± 0.02 ^{k}	0.62 ± 0.60 ^{ab}	0.57 ± 0.17 ^{bc}	2.48 ± 0.25 ^{l}
DGO-5RW	9.33 ± 0.05 ^{hi}	0.95 ± 0.02 ^{a}	1.05 ± 0.03 ^{a}	0.39 ± 0.02 ^{b}	1.04 ± 0.01 ^{bcd}	0.96 ± 0.01 ^{b}	3.42 ± 0.02 ^{j}	0.66 ± 0.15 ^{abc}	0.65 ± 0.18 ^{bcde}	2.45 ± 0.19 ^{l}
DHO-5RW	8.95 ± 0.07 ^{f}	1.60 ± 0.00 ^{e}	1.06 ± 0.01 ^{a}	0.40 ± 0.00 ^{bc}	1.06 ± 0.00 ^{bcde}	1.58 ± 0.12 ^{f}	2.87 ± 0.01 ^{de}	0.66 ± 0.12 ^{abc}	0.65 ± 0.00 ^{bcde}	1.28 ± 0.01 ^{e}
DOL-5RW	3.70 ± 0.10 ^{b}	1.20 ± 0.10 ^{b}	1.13 ± 0.01 ^{b}	0.31 ± 0.01 ^{a}	1.03 ± 0.00 ^{bcd}	1.87 ± 0.01 ^{k}	4.15 ± 0.01 ^{m}	0.83 ± 0.02 ^{d}	0.72 ± 0.08 ^{e}	2.27 ± 0.01 ^{k}
DSO-5RW	9.66 ± 0.05 ^{jk}	1.50 ± 0.02 ^{efg}	1.09 ± 0.03 ^{ab}	0.44 ± 0.02 ^{bcdef}	1.12 ± 0.02 ^{cde}	0.98 ± 0.00 ^{b}	2.85 ± 0.01 ^{d}	0.65 ± 0.53 ^{abc}	0.67 ± 0.13 ^{cde}	1.86 ± 0.01 ^{h}
DWO-5RW	9.40 ± 0.10 ^{hij}	1.45 ± 0.02 ^{cde}	1.07 ± 0.02 ^{ab}	0.43 ± 0.00 ^{bcdef}	1.13 ± 0.01 ^{de}	1.70 ± 0.01 ^{hi}	2.95 ± 0.01 ^{i}	0.64 ± 0.20 ^{abc}	0.70 ± 0.01 ^{e}	1.24 ± 0.01 ^{de}
DGO-9RW	10.93 ± 0.15 ^{l}	5.72 ± 0.03 ^{j}	1.08 ± 0.03 ^{ab}	0.46 ± 0.01 ^{efg}	0.90 ± 0.01 ^{a}	0.86 ± 0.012 ^{a}	4.29 ± 0.01 ^{n}	0.62 ± 0.27 ^{ab}	0.43 ± 0.10 ^{a}	3.43 ± 0.01 ^{m}
DHO-9RW	7.40 ± 0.10 ^{d}	5.73 ± 0.02 ^{j}	1.10 ± 0.00 ^{ab}	0.47 ± 0.00 ^{fg}	2.46 ± 0.02 ^{i}	2.38 ± 0.33 ^{n}	3.59 ± 0.01 ^{l}	0.64 ± 0.01 ^{abc}	1.99 ± 0.00 ^{h}	1.20 ± 0.01 ^{c}
DOL-9RW	9.83 ± 0.05 ^{k}	1.35 ± 0.01 ^{c}	1.05 ± 0.01 ^{a}	0.43 ± 0.00 ^{bcdef}	1.01 ± 0.00 ^{ab}	0.97 ± 0.01 ^{b}	2.80 ± 0.01 ^{c}	0.63 ± 0.01 ^{abc}	0.57 ± 0.01 ^{bcd}	1.83 ± 0.01 ^{gh}
DSO-9RW	9.66 ± 0.15 ^{k}	1.50 ± 0.02 ^{efg}	1.07 ± 0.01 ^{ab}	0.44 ± 0.01 ^{bcdef}	1.03 ± 0.01 ^{bcd}	0.98 ± 0.01 ^{b}	2.85 ± 0.00 ^{d}	0.63 ± 0.01 ^{abc}	0.58 ± 0.01 ^{bcd}	1.87 ± 0.01 ^{h}
DWO-9RW	2.66 ± 0.05 ^{a}	0.93 ± 0.02 ^{a}	1.08 ± 0.03 ^{a}	0.42 ± 0.00 ^{bcde}	1.10 ± 0.01 ^{bcde}	1.67 ± 0.01 ^{gh}	2.89 ± 0.01 ^{efg}	0.64 ± 0.01 ^{abc}	0.68 ± 0.01 ^{de}	1.21 ± 0.01 ^{cd}

Mixolab parameters: ST, stability; DDT, dough development time; C1–C5, maximum consistency during stages 1–5, respectively; C12, difference between torques C1 and C2; C32, difference between torques C3 and C2; C34, difference between torques C3 and C4; C54, difference between torques C5 and C4. Samples: WF— wheat flour; WF-5SG—wheat flour with 5% shortening addition; DGO-5BW, DGO-9BW DGO-5RW, and DGO-9RW—dough samples with 5% addition of oleogel with grape seed oil and 5% and 9% beeswax and, respectively, rice bran wax; DHO-5BW, DHO-9BW, DHO-5RW, and DHO-9RW—dough samples with 5% addition of oleogel with hemp seed oil and 5% and 9% beeswax and, respectively, rice bran wax; DOL-5BW, DOL-9BW, DOL-5RW, and DOL-9RW—dough samples with 5% addition of oleogel with olive oil and 5% and 9% beeswax and, respectively, rice bran wax; DSO-5BW, DSO-9BW, DSO-5RW, and DSO-9RW—dough samples with 5% addition of oleogel with sunflower oil and 5% and 9% beeswax and, respectively, rice bran wax; DWO-5BW, DWO-9BW, DWO-5RW, and DWO-9RW—dough samples with 5% addition of oleogel with walnut oil and 5% and 9% beeswax and, respectively, rice bran wax. Means in the same column with different letters for each sample type are significantly different ($p < 0.05$).

According to Baltsavias et al. [47], the hard fat may be spread to giant lumps during mixing, whereas the soft one is fragmented around the flour particles. Oils are less powerful in their oleogel system and may be dispersed during the mixing of all of them through the dough in the form of minuscule globules, thereby reducing dough consistency. The behavior of the dough samples during the gelatinization process changes significantly ($p < 0.05$) by adding different types of oleogels into wheat flour. It is generally accepted that in the presence of lipids when starch gelatinization takes place, the starch-amylose complexes have lower viscosity [45]. Molecular interaction takes place between amylose and the oleogels, thereby decreasing the leaching of amylose and reducing the swelling of starch, possibly leading to a reduction in the dough starch parameter values. This decreased all the Mixolab parameters related to the starch characteristics such as starch gelatinization (C3 and C32), cooking stability (C4), and starch gelling (C5 and C54). Through the oleogel or shortening addition in dough, some complexes between lipids and wheat flour compound protein and starch may be formed. This prevents the leaching of amylase from starch when it begins to gelatinize. Therefore, the swelling of the starch granules is delayed and the starch gelatinization process inhibited [48]. During the baking process, through the energy input, the dynamic equilibrium state changes, and the dough turns into bread through the starch gelatinization process and gluten denaturation network. Through the gelatinization process, starch changes its conformation, free glucan chains appear, and hydrophobic areas and hydrophilic surfaces change. Proteins occur as well as irreversible structural changes through thermal denaturation [44]. Under these conditions, there is a translocation of the lipids in the dough system. The lipids migrate from the gluten complex to the partially gelatinized starch, forming complexes with their main compound's amylopectin and amylose, thus slowing down starch retrogradation after baking. This slowdown is reflected by an increase in bread shelf life indicated by the Mixolab decrease value of C54 due to the oleogel addition. The delay of the starch retrogradation process due to the addition of oleogels in the wheat flour dough, which prolongs the shelf life of bakery products, has also been reported by others [40,41,49].

2.8. Texture Profile Analysis of the Wheat Flour Dough with Different Types of Oleogel Additions

The textural parameters of the wheat flour dough with different types of oleogel additions at a 5% addition level in wheat flour are shown in Table 6. Significant differences ($p < 0.05$) in dough hardness and elasticity were obtained in samples with beeswax oleogels, especially in those with olive oil oleogels (DOL-9BW) and hemp seed oil oleogels (DHO-9BW) obtained with 9% wax. The results are in accordance with other works showing that the use of oleogels significantly affects the texture of the baked product, especially the hardness [48–50] and the elasticity of the final product, when oleogels are used as substitutes for traditional lipids. If the hardness value increases for doughs with an oleogel obtained with 9% wax, then the adhesiveness decreases significantly ($p < 0.05$). This behavior may be due to a higher level of SFAs in these types of oleogels that are fully saturated with hydrogen and present a more rigid and fixed structure being more solid-like and compact-packed in the dough system [25]. The doughs with the addition of 5% oleogel in its recipe were less sticky and were easier to process in a round format, specifically for buns. Cohesiveness was inversely proportional to adhesiveness. The dough samples with low adhesiveness (DGO-9BW; DHO-9BW) showed high cohesiveness values. Cohesion quantifies the internal strength of the food structure and provides information about the ability of a material to stick to itself [51]. Therefore, the oleogels analyzed in this study could be used in the fermented dough to improve the elasticity and hardness properties if they are formulated with a maximum of 5% added wax. Wax in a percentage of 9% used in the formulation of oleogels can lead to doughs with high hardness values.

Table 6. TPA profile analysis of dough with oleogels.

Sample	Hardness (g)	Stickiness (%)	Adhesiveness (g.s)	Cohesiveness (%)	Springiness (%)
DGO_5BW	772 [j]	−127.4 [b]	−1294 [k]	38.4 [e]	1 [NS]
DHO_5BW	729 [p]	−216 [gh]	−1453 [l]	27.6 [i]	1 [NS]
DOL_5BW	845 [d]	−203 [e]	−2145 [p]	34.2 [f]	1 [NS]
DSO_5BW	757 [m]	−128 [b]	−1876 [n]	33.7 [fg]	1 [NS]
DWO_5BW	763 [l]	−212 [fg]	−1022 [j]	40.1 [de]	1 [NS]
DGO_9BW	886 [c]	−124 [ab]	−890 [g]	54.1 [a]	1 [NS]
DHO_9BW	954 [a]	−236 [ij]	−789 [d]	56.8 [a]	1 [NS]
DOL_9BW	947 [b]	−236 [j]	−875 [f]	47.5 [b]	1 [NS]
DSO_9BW	821 [e]	−169 [c]	−987 [i]	40.2 [de]	1 [NS]
DWO_9BW	802 [l]	−209 [ef]	−908 [h]	40.8 [de]	1 [NS]
DGO_5RW	766 [k]	−189 [d]	−2765 [r]	33.4 [fg]	1 [NS]
DHO_5RW	751 [n]	−190 [d]	−2431 [q]	30.7 [ghi]	1 [NS]
DOL_5RW	804 [g]	−207 [ef]	−1899 [o]	31.5 [fgh]	1 [NS]
DSO_5RW	741 [o]	−221 [h]	−1765 [m]	29.9 [hi]	1 [NS]
DWO_5RW	75 2[n]	−216 [gh]	−2908 [s]	27.6 [i]	1 [NS]
DGO_9RW	780 [l]	−119 [a]	−679 [b]	38.3 [e]	1 [NS]
DHO_9RW	759 [lm]	−125 [ab]	−709 [c]	38.7[de]	1 [NS]
DOL_9RW	814 [f]	−208 [ef]	−814 [e]	40.3 [de]	1 [NS]
DSO_9RW	847 [d]	−230 [i]	−712 [c]	44.3 [bc]	1[NS]
DWO_9RW	794 [h]	−218 [gh]	−608 [a]	41.9 [cd]	1 [NS]
F-value	12,313.03 ***	1301.50 ***	1622 × 10^3 ***	190.65 ***	

DGO_5BW, DGO_9BW DGO_5RW, and DGO_9RW—dough samples with 5% addition of oleogel with grape seed oil and 5% and 9% beeswax and, respectively, rice bran wax; DHO_5BW, DHO_9BW, DHO_5RW, and DHO_9RW—dough samples with 5% addition of oleogel with hemp seed oil and 5% and 9% beeswax and, respectively, rice bran wax; DOL_5BW, DOL_9BW, DOL_5RW, and DOL_9RW—dough samples with 5% addition of oleogel with olive oil and 5% and 9% beeswax and, respectively, rice bran wax; DSO_5BW, DSO_9BW, DSO_5RW, and DSO_9RW—dough samples with 5% addition of oleogel with sunflower oil and 5% and 9% beeswax and, respectively, rice bran wax; DWO_5BW, DWO_9BW, DWO_5RW, and DWO_9RW—dough samples with 5% addition of oleogel with walnut oil and 5% and 9% beeswax and, respectively, rice bran wax. Means in the same column with different letters for each sample type are significantly different ($p < 0.05$), and the F-value represents Tukey's HSD test, *** strong correlation. NS—insignificant.

2.9. Principal Component Analysis of Thermal and Rheological Properties of Oleogels

The graphical representation of the principal component analysis is shown in Figure 5. In this graph, a correspondence of the midpoint, enthalpy variables (DSC), and the transition temperature of the oleogels from the solid phase to the liquid phase (rheological properties) has been made. The graph of the main component's groups for the oleogel samples according to the rheological properties and the transition point from the gel phase to the liquid phase has been made. According to the data obtained, the oleogels obtained with beeswax have a lower melting point compared with the oleogels obtained with rice bran wax. The inclusion of these oleogels in separate quadrants depends on the oleogelator used and the rheological properties of the oleogels. The oleogels with 5% rice bran wax correlate with the transition temperature from the gel to the liquid phase. The exception is the sunflower oleogel (SO) obtained with rice bran wax (SO-5RW) and the one with beeswax (SO-5BW), which do not have similar characteristics to the oleogels from the same group.

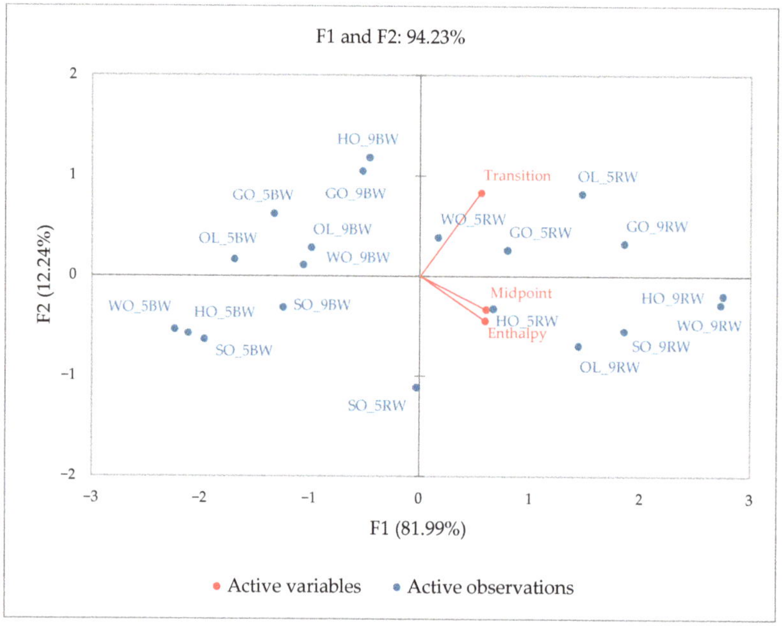

Figure 5. PCA graph for thermal and rheological properties of oleogels.

3. Conclusions

The oleogels used in the current study prepared on the base of the two wax assortments of BW and RW and five types of vegetable oils have the potential to be an alternative to unhealthy fats that are usually used in bakery industry. All the oleogels analyzed presented good quality characteristics for use in the food industry. The melting peak was recorded at a lower temperature for beeswax oleogels compared with that of the oleogels that contain rice bran wax. All the oleogel samples presented a solid, elastic-like behavior. The oleogel samples obtained with grape seed oil had higher viscoelasticity, followed by oleogels with hemp seed oil and oleogels with walnut oil. Regarding the dough rheological behavior, according to the data, a solid, elastic-like one was observed for all the samples across the whole range of frequencies. The addition of oleogels from olive oil and hemp seed oil in wheat flour dough led to high values for the G′ and G″ moduli, whereas the addition of oleogels based on walnut oil led to low values. Thermogravimetric analysis of oleogels described the decomposition of oleogels in two distinct stages as follows: the first stage corresponded to the decomposition of short-chain fatty and unsaturated acids at a temperature between 295 °C and 480 °C, and the second stage in the temperature range of 480–550 °C was associated with the thermal volatilization of the vegetable oil used to formulate oleogels. From the point of view of dough's rheological properties during fermentation, doughs with rice bran wax presented better fermentation rheological values. The retention coefficient presented higher values for all dough samples, which indicated that the loaf volume of buns is adequate for all the samples with an addition of oleogels in their recipe. However, the use of oleogels based on walnut and hemp seed oil presented lower values for the CR parameter that indicated a lower capacity to retain gasses in the dough system when these types of oleogels were added. The results obtained with the Mixolab device indicated that the addition of oleogels significantly influenced the thermo-mechanical behavior of the wheat flour doughs. The use of oleogels in the dough system ensured a better stability during mixing, a decrease in the dough viscosity during heating and cooling, a delay of the starch gelatinization process, and better cooking stability and starch gelling. Our data indicated that the oleogels can improve the shelf life of the bakery products due to the fact that the difference between torques C5 and C4 decreased in the dough samples with oleogels in their recipe when compared with the control sample.

4. Materials and Methods

4.1. Materials

Beeswax and rice bran wax were supplied by Merck Romania. The vegetable oils of grape seed oil (GO), hemp seed oil (HO), olive oil (OL), sunflower oil (SO), and walnut oil (WO) were purchased from grocery stores. Wheat flour, type 650, was purchased from a local supplier. Yeast and salt were purchased from the local supermarket (Suceava, Romania).

4.2. Preparation of Oleogels

Oleogels were obtained using vegetable oils from 5 sources (olive, walnut, sunflower, grape seed, and hemp). Yellow beeswax and rice bran wax were used as oleogelation agents in percentages of 5% and 9%. These percentages have been chosen according to the literature where the main range of the level of the structure-forming agent in oleogels is from 0.5% *w*/*w* to 10.0% *w*/*w* [52,53]. For beeswax, the critical concentration of the oleogel formation for most vegetable oils is around 6.0% *w*/*w* [17], whereas for rice bran it is 5.0% *w*/*w* [10]. The oleogels were obtained according to the method described by Ropciuc et al. [3]. The oils and wax were heated to a temperature of 80 °C using a hotplate with an orbital stirrer until the wax completely dissolved. A total of 20 samples of oleogels were obtained for determinations. The mixture obtained was poured into tubes for solidification. The oleogels were stored under refrigerated conditions at 4 °C for five days until analysis. The samples were coded according to the type of oil used and the percentage of wax added to obtain the oleogel as follows: GO_5BW (oleogel from grape seed oil and 5% beeswax), HO_5BW (oleogel from hemp seed oil and 5% beeswax), OL_5BW (oleogel from olive oil and 5% beeswax), SO_5BW (oleogel from sunflower oil and 5% beeswax), and WO_5% (oleogel from walnut oil and 5% beeswax). The same form of coding was used for oleogels with a percentage of 9% beeswax. Rice bran wax samples were coded using the RW term (e.g., GO_5RW) [42,50,54].

4.3. Dough Preparation for Fundamental Rheological Analysis

Dough was obtained using flour type 650, salt, and water, into which 5% oleogel was incorporated according to the recipe described in Table 7. The hydration capacity of the flour was determined using the simulation test of the Mixolab device.

Table 7. Dough components for rheological analysis.

Dough Components for Fundamental Rheological Determination				Dough Components for Rheological Property Analysis during Fermentation				
Flour, g	**Salt, g**	**Water, mL**	**Oleogel, g**	**Flour, g**	**Salt, g**	**Yeast, g**	**Water, mL**	**Oleogel, g**
30	0.45	17.4	1.5	300	5	3	174	15

4.4. Dough Preparation for Rheological Properties Analysis during Fermentation

The dough samples for rheological property analysis during fermentation were obtained using type 650 flour, salt, yeast, water, and 5% oleogel.

4.5. Rheological Behavior of Oleogels and Dough

All rheological measurements were performed using a Haake Mars 40 Dynamic Rheometer (Karlsruhe, Germany) equipped with a Peltier system and circulating water bath (Huber, Germany) for temperature control. A glossy plate rotor with a diameter of 40 mm was used, with the distance between the plates being 1 mm. The linear viscoelastic region (LVR) of the oleogels was determined by logarithmic growth with the oscillation stress being from 0.1 to 100 Pa at a frequency of 1.0 Hz. The frequency sweep range was set from 0.1 to 10 Hz at a temperature of 25 °C. The viscoelastic properties of the oleogel samples were also evaluated under the influence of temperature. Modulus G′, "elastic" modulus G″, and "loss" were analyzed at temperatures between 20 and 90 °C, with a temperature increase of 2 °C/min according to the method previously described by Pușcaș et al. [4].

4.6. Differential Scanning Calorimetry

The influence of oleogel composition on thermal behavior was studied by differential scanning calorimetry (DSC). A DSC 25 calorimeter (TA Instruments, New Castle, DE, USA) calibrated with indium (28.71 J/g) at 156.6 °C was used for the analysis. An amount of approximately 4 mg for each sample was weighed and hermetically sealed in an aluminum pan that was inserted into the analysis instrument along with an empty pan used as a reference. The samples were first equilibrated at 20 °C, and then the temperature was varied in three steps as follows: from 20 °C to 90 °C, from 90 °C to −60 °C, and from −60 °C to 100 °C. The temperature was finally lowered to 20 °C. The heating rate was 10 °C/min and nitrogen was used as a purge gas at a flow rate of 20 mL/min.

4.7. Thermogravimetric Analysis of Oleogels

Thermogravimetric stability of the oleogels was evaluated using a thermal analysis system TGA 2 (Mettler Toledo, Columbus, OH, USA) by placing 5–10 mg of the sample in an aluminum oxide pan and subjecting it to heating from 30 to 600 °C at a heating rate of 10 °C/min in a nitrogen atmosphere (20 mL/min) according to the method previously described by Alvarez-Ramirez [55].

4.8. Texture Profile Analysis of Oleogels

To evaluate the textural properties of the oleogel samples with different oleogelation agents (beeswax and rice bran wax), a TVT-6700 texturometer (Perten Instruments, Hägersten, Sweden) was used with an aluminum cylinder of 25 mm diameter. The oleogel samples with a mass of 4 g were placed on a plastic plate with a diameter of 28 mm. The double compression method was used, with a compression percentage of 30%. The trigger force used was 10 g, the velocity was 5.0 mm/s, and the recovery period between compressions was 10 s.

4.9. Rheological Properties of Dough during Fermentation

The rheological properties of the dough during fermentation were determined using the Chopin Rheofermentometer F4 (Chopin, Villeneuve la Garenne, France) according to the Chopin protocol. The method used has been previously described by Atudorei et al. [56]. For this purpose, 315 g of dough was introduced into the curve of the Rheofermentometer, and it was pressed on top with a weight of 2 kg and subjected to fermentation at a temperature of 30 °C for 3 h. The parameters of dough formation, the maximum level of gas production (H'm), the volume of the gas retained in the dough at the end of the test (VR), the total volume of CO_2 production (VT), and the retention coefficient (CR) were determined.

4.10. Rheological Characteristics of Dough Using Mixolab Device

The effect of different types of oleogel addition in wheat flour on the thermo-mechanical characteristics of dough was made using a Mixolab 2 device (Chopin SA., Villeneuve la Garenne, France) according to ICC Standard Method No. 173 [57]. Wheat flour with 5% of the different types of oleogel additions was mixed into the Mixolab bowl in which distilled water was added to obtain the optimum dough consistency. The protocol used for Mixolab analysis was the following: mixer temperature at 30 °C, kneading speed at 80 rpm, heating rate of 4 °C/min to 90 °C, and cooling rate of 4 °C/min to 50 °C.

4.11. Dough Texture Analysis

Dough texture analysis was evaluated using a TVT-6700 texture analyzer (Perten Instruments, Hägersten, Sweden) according to the method previously described by Jung et al. [41]. A dough with a mass of 50 g (according to the recipe presented in Table 7—dough components for rheological properties analysis during fermentation) was shaped into a sphere and placed on the platform, and a cylindrical compression platen probe of 35 mm diameter was used to perform a TPA test. The sample was compressed twice at a speed of 5 mm/s for 25 s followed by a rest time of 12 s between the two cycles and 10 mm/s retracting speed at a data rate of 200 pps. The textural parameters determined were hardness, stickiness, adhesiveness, cohesiveness, and springiness.

4.12. Statistical Analysis

Data reported in this study are mean values, and significant differences between samples were analyzed by one-way analysis of variance (ANOVA) by Tukey's test ($p < 0.05$) using XLSTAT software, ver. 2023 (Addinsoft, New York, NY, USA). Pearson correlation (r) and principal component analysis (PCA) were also performed to determine relationships between different properties using XLSTAT software, ver. 2023.

Supplementary Materials: The following supporting information can be downloaded at: https://www.mdpi.com/article/10.3390/gels10030194/s1. Table S1: DSC parameters of oleogel samples, oils, and waxes—second heating and second cooling profile; Figure S1: Thermograms of oleogels, oils, and waxes for heating (→) and cooling (←) profiles; Figure S2: TGA (dashed line)/DTG (solid line) thermograms for oleogels formulated with different types of oil and wax.

Author Contributions: Conceptualization, S.R., F.D. and G.G.C.; methodology, S.R., G.G.C. and F.D.; software, S.R.; validation, S.R., G.G.C. and M.A.O.; formal analysis, A.E.P., A.L., S.R., F.D. and M.S.; investigation, S.R., A.L., A.E.P. and G.G.C.; resources, G.G.C., M.A.O., A.L. and F.D.; data curation, S.R.; writing—original draft preparation, S.R., F.D. and G.G.C.; writing—review and editing, S.R., A.L. and F.D.; visualization, S.R. and G.G.C.; supervision, F.D. and S.R.; project administration, S.R.; funding acquisition, S.R. All authors have read and agreed to the published version of the manuscript.

Funding: This work was supported by a grant from the Ministry of Research, Innovation, and Digitization, CNCS-UEFISCDI, project number PN-III-P2-2.1-PED-2021-1738, within PNCDI III.

Institutional Review Board Statement: Not applicable.

Informed Consent Statement: Not applicable.

Data Availability Statement: The original contributions presented in the study are included in the article/Supplementary Materials, further inquiries can be directed to the corresponding authors.

Acknowledgments: This work was supported by a grant from the Ministry of Research, Innovation and Digitization, CNCS-UEFISCDI, project number PN-III-P2-2.1-PED-2021-1738, within PNCDI III.

Conflicts of Interest: The authors declare no conflicts of interest. Informed consent was obtained from all subjects involved in the study.

References

1. Pipoyan, D.; Stepanyan, S.; Stepanyan, S.; Beglaryan, M.; Costantini, L.; Molinari, R.; Merendino, N. The effect of trans fatty acids on human health: Regulation and consumption patterns. *Foods* **2021**, *10*, 2452. [CrossRef]
2. Demirkesen, I.; Mert, B. Recent developments of oleogel utilizations in bakery products. *Crit. Rev. Food Sci. Nutr.* **2020**, *60*, 2460–2479. [CrossRef]
3. Ropciuc, S.; Dranca, F.; Oroian, M.A.; Leahu, A.; Codină, G.G.; Prisacaru, A.E. Structuring of cold pressed oils: Evaluation of the physicochemical characteristics and microstructure of white beeswax oleogels. *Gels* **2023**, *9*, 216. [CrossRef] [PubMed]
4. Pușcaș, A.; Tanislav, A.E.; Mureșan, A.E.; Fărcaș, A.C.; Mureșan, V. Walnut oil oleogels as milk fat replacing system for commercially available chocolate butter. *Gels* **2022**, *8*, 613. [CrossRef] [PubMed]
5. Uzunova, G.; Perifanova-Nemska, M.; Stojanova, M.; Gandev, S. Chemical composition of walnut oil from fruits on different years old branches. *Bulg. J. Agric. Sci.* **2015**, *21*, 494–497.
6. Petraru, A.; Ursachi, F.; Amariei, S. Nutritional characteristics assessment of sunflower seeds, oil and cake. Perspective of using sunflower oilcakes as a functional ingredient. *Plants* **2021**, *10*, 2487. [CrossRef] [PubMed]

7. Wijarnprecha, K.; Aryusuk, K.; Santiwattana, P.; Sonwai, S.; Rousseau, D. Structure and rheology of oleogels made from rice bran wax and rice bran oil. *Food Res. Int.* **2018**, *112*, 199–208. [CrossRef] [PubMed]
8. Winkler-Moser, J.K.; Anderson, J.; Felker, F.C.; Hwang, H. Physical properties of beeswax, sunflower Wax, and candelilla wax mixtures and oleogels. *J. Am. Oil Chem. Soc.* **2019**, *96*, 1125–1142. [CrossRef]
9. Dimakopoulou-Papazoglou, D.; Giannakaki, F.; Katsanidis, E. Structural and physical characteristics of mixed-component oleogels: Natural wax and monoglyceride interactions in different edible oils. *Gels* **2023**, *9*, 627. [CrossRef]
10. Pandolsook, S.; Kupongsak, S. Influence of bleached rice bran wax on the physicochemical properties of organogels and water-in-oil emulsions. *J. Food Eng.* **2017**, *214*, 182–192. [CrossRef]
11. Modupalli, N.; Thangaraju, S.; Naik, G.M.; Rawson, A.; Natarajan, V. Assessment of physicochemical, functional, thermal, and phytochemical characteristics of refined rice bran wax. *Food Chem.* **2022**, *396*, 133737. [CrossRef]
12. Pinto, T.C.; Martins, A.J.; Pastrana, L.; Pereira, M.C.; Cerqueira, M.A. Oleogel-based systems for the delivery of bioactive compounds in foods. *Gels* **2021**, *7*, 86. [CrossRef]
13. Mert, B.; Demirkesen, I. Evaluation of highly unsaturated oleogels as shortening replacer in a short dough product. *LWT Food Sci. Technol.* **2016**, *68*, 477–484. [CrossRef]
14. Mert, B.; Demirkesen, I. Reducing saturated fat with oleogel/shortening blends in a baked product. *Food Chem.* **2016**, *199*, 809–816. [CrossRef] [PubMed]
15. Doan, C.D.; To, C.M.; De Vrieze, M.; Lynen, F.; Danthine, S.; Brown, A.; Dewettinck, K.; Patel, A.R. Chemical profiling of the major components in natural waxes to elucidate their role in liquid oil structuring. *Food Chem.* **2017**, *214*, 717–725. [CrossRef]
16. Barragán-Martínez, L.P.; Román-Guerrero, A.; Vernon-Carter, E.J.; Alvarez-Ramirez, J. Impact of fat replacement by a hybrid gel (canola oil/candelilla wax oleogel and gelatinized corn starch hydrogel) on dough viscoelasticity, color, texture, structure, and starch digestibility of sugar-snap cookies. *Int. J. Gastron. Food Sci.* **2022**, *29*, 100563. [CrossRef]
17. Frolova, Y.; Sarkisyan, V.; Sobolev, R.; Makarenko, M.; Semin, M.; Kochetkova, A. The influence of edible oils' composition on the properties of beeswax-based oleogels. *Gels* **2022**, *8*, 48. [CrossRef] [PubMed]
18. Tanislav, A.E.; Pușcaș, A.; Păucean, A.; Mureșan, A.E.; Semeniuc, C.A.; Mureșan, V.; Mudura, E. Evaluation of structural behavior in the process dynamics of oleogel-based tender dough products. *Gels* **2022**, *8*, 317. [CrossRef]
19. Blake, A. The Microstructure and Physical Properties of Plant-Based Waxes and Their Relationship to the Oil Binding Capacity of Wax Oleogels. Doctoral Dissertation, University of Guelph, Guelph, ON, Canada, 2015.
20. Tavernier, I.; Doan, C.D.; Van de Walle, D.; Danthine, S.; Rimaux, T.; Dewettinck, K. Sequential crystallization of high and low melting waxes to improve oil structuring in wax-based oleogels. *RSC Adv.* **2017**, *7*, 12113–12125. [CrossRef]
21. Thakur, D.; Singh, A.; Prabhakar, P.K.; Meghwal, M.; Upadhyay, A. Optimization and characterization of soybean oil-carnauba wax oleogel. *LWT* **2022**, *157*, 113108. [CrossRef]
22. Patel, A.R.; Babaahmadi, M.; Lesaffer, A.; Dewettinck, K. Rheological profiling of organogels prepared at critical gelling concentrations of natural waxes in atriacylglycerol solvent. *J. Agric. Food Chem.* **2015**, *63*, 4862–4869. [CrossRef] [PubMed]
23. Badem, Ş.; Baştürk, A. Oxidative stability and characterization of oleogels obtained from safflower oil-based beeswax and rice bran wax and their effect on the quality of cake samples. *J. Am. Oil Chem. Soc.* **2023**, *100*, 635–649. [CrossRef]
24. Gutiérrez-Luna, K.; Astiasaran, I.; Ansorena, D. Fat reduced cookies using an olive oil-alginate gelled emulsion: Sensory properties, storage stability and in vitro digestion. *Food Res. Int.* **2023**, *167*, 112714. [CrossRef] [PubMed]
25. Zhang, T.; Guan, E.; Yang, Y.; Liu, F.; Zhang, L.; Pang, J.; Bian, K. Fatty acid profiles of vegetable oils from four different plant sources and their effects on dough rheology and Chinese steamed bread quality. *Int. J. Food Sci. Technol.* **2021**, *56*, 2407–2414. [CrossRef]
26. Espert, M.; Wang, Q.; Sanz, T.; Salvador, A. Sunflower oil-based oleogel as fat replacer in croissants: Textural and sensory characterisation. *Food Bioprocess Technol.* **2023**, *16*, 1943–1952. [CrossRef]
27. Zhao, M.; Lan, Y.; Cui, L.; Monono, E.; Rao, J.; Chen, B. Formation, characterization, and potential food application of rice bran wax oleogels: Expeller-pressed corn germ oil versus refined corn oil. *Food Chem.* **2020**, *309*, 125704. [CrossRef]
28. Doan, C.D.; Tavernier, I.; Danthine, S.; Rimaux, T.; Dewettinck, K. Physical compatibility between wax esters and triglycerides in hybrid shortenings and margarines prepared in rice bran oil. *J. Sci. Food Agric.* **2018**, *98*, 1042–1051. [CrossRef]
29. Wei, F.; Miao, J.; Tan, H.; Feng, R.; Zheng, Q.; Cao, Y.; Lan, Y. Oleogel-structured emulsion for enhanced oxidative stability of perilla oil: Influence of crystal morphology and cooling temperature. *LWT* **2021**, *139*, 110560. [CrossRef]
30. Espinosa-Solís, V.; García-Tejeda, Y.V.; Portilla-Rivera, O.M.; Barrera-Figueroa, V. Tailoring olive oil microcapsules via microfluidization of pickering o/w emulsions. *Food Bioprocess Technol.* **2021**, *14*, 1835–1843. [CrossRef]
31. Pereira, L.H.; Pereira, J.; Garcia, J.S.; Trevisan, M.G. Seed oil detection in extra virgin olive oil by differential scanning calorimetry. *J. Therm. Anal. Calorim.* **2023**, *148*, 6833–6843. [CrossRef]
32. Gila, A.; Sánchez-Ortiz, A.; Jiménez, A.; Beltrán, G. The ultrasound application does not affect to the thermal properties and chemical composition of virgin olive oils. *Ultrason. Sonochem.* **2021**, *70*, 105320. [CrossRef]
33. Wirkowska-Wojdyła, M.; Ostrowska-Ligęza, E.; Górska, A.; Bryś, J. Application of chromatographic and thermal methods to study fatty acids composition and positional distribution, oxidation kinetic parameters and melting profile as important factors characterizing amaranth and quinoa oils. *Appl. Sci.* **2022**, *12*, 2166. [CrossRef]
34. Sánchez, R.; Alonso, G.; Valencia, C.; Franco, J.M. Rheological and TGA study of acylated chitosan gel-like dispersions in castor oil: Influence of acyl substituent and acylation protocol. *Chem. Eng. Res. Des.* **2015**, *100*, 170–178. [CrossRef]

35. Çokay, H.; Öğütcü, M. Determining the structure and stability of essential oil-sunflower wax and beeswax oleogels. *J. Am. Oil Chem. Soc.* **2023**, *100*, 993–1002. [CrossRef]
36. Martín-Alfonso, M.A.; Rubio-Valle, J.F.; Hinestroza, J.P.; Martín-Alfonso, J.E. Impact of Vegetable Oil Type on the Rheological and Tribological Behavior of Montmorillonite-Based Oleogels. *Gels* **2022**, *8*, 504. [CrossRef] [PubMed]
37. Bucio, A.; Moreno-Tovar, R.; Bucio, L.; Espinosa-Dávila, J.; Anguebes-Franceschi, F. Characterization of Beeswax, Candelilla Wax and Paraffin Wax for Coating Cheeses. *Coatings* **2021**, *11*, 261. [CrossRef]
38. Öğütcü, M.; Yılmaz, E. Characterization of hazelnut oil oleogels prepared with sunflower and carnauba waxes. *Int. J. Food Prop.* **2015**, *18*, 1741–1755. [CrossRef]
39. Dassanayake, L.S.K.; Kodali, D.R.; Ueno, S.; Sato, K. Physical properties of rice bran wax in bulk and organogels. *J. Am. Oil Chem. Soc.* **2009**, *86*, 1163. [CrossRef]
40. Bascuas, S.; Morell, P.; Quiles, A.; Salvador, A.; Hernando, I. Use of oleogels to replace margarine in steamed and baked buns. *Foods* **2021**, *10*, 1781. [CrossRef]
41. Jung, D.; Oh, I.; Lee, J.; Lee, S. Utilization of butter and oleogel blends in sweet pan bread for saturated fat reduction: Dough rheology and baking performance. *LWT* **2020**, *125*, 109194. [CrossRef]
42. Marangoni, A.G. Organogels: An alternative edible oil-structuring method. *J. Am. Oil Chem. Soc.* **2012**, *89*, 749–780. [CrossRef]
43. Mironeasa, S.; Codină, G.G.; Mironeasa, C. The effects of wheat flour substitution with grape seed flour on the rheological parameters of the dough assessed by Mixolab. *J. Texture Stud.* **2012**, *43*, 40–48. [CrossRef]
44. Pareyt, B.; Finnie, S.M.; Putseys, J.A.; Delcour, J.A. Lipids in bread making: Sources, interactions, and impact on bread quality. *J. Cereal Sci.* **2011**, *54*, 266–279. [CrossRef]
45. Banu, I.; Patrașcu, L.; Vasilean, I.; Dumitrașcu, L.; Aprodu, I. Influence of the protein-based emulsions on the rheological, thermo-mechanical and baking performance of muffin formulations. *Appl. Sci.* **2023**, *13*, 3316. [CrossRef]
46. Oh, I.K.; Amoah, C.; Lim, J.; Jeong, S.; Lee, S. Assessing the effectiveness of wax-based sunflower oil oleogels in cakes as a shortening replacer. *LWT* **2017**, *86*, 430–437. [CrossRef]
47. Baltsavias, A.; Jurgens, A.; van Vliet, T. Rheological properties of short doughs at small deformation. *J. Cereal Sci.* **1997**, *26*, 289–300. [CrossRef]
48. Codină, G.G.; Mironeasa, S. Use of response surface methodology to investigate the effects of brown and golden flaxseed on wheat flour dough microstructure and rheological properties. *J. Food Sci. Technol.* **2016**, *53*, 4149–4158. [CrossRef] [PubMed]
49. Malvano, F.; Laudisio, M.; Albanese, D.; D'Amore, M.; Marra, F. Olive oil-based oleogel as fat replacer in a sponge cake: A comparative study and optimization. *Foods* **2022**, *11*, 2643. [CrossRef]
50. Kim, J.Y.; Lim, J.; Lee, J.; Hwang, H.; Lee, S. Utilization of oleogels as a replacement for solid fat in aerated baked goods: Physicochemical, rheological, and tomographic characterization. *J. Food Sci.* **2017**, *82*, 445–452. [CrossRef]
51. Lu, T.-M.; Lee, C.-C.; Mau, J.-L.; Lin, S.-D. Quality and antioxidant property of green tea sponge cake. *Food Chem.* **2010**, *119*, 1090–1095. [CrossRef]
52. Doan, C.D.; Van de Walle, D.; Dewettinck, K.; Patel, A.R. Evaluating the oil-gelling properties of natural waxes in rice bran oil: Rheological, thermal, and microstructural study. *J. Am. Oil Chem. Soc.* **2015**, *92*, 801–811. [CrossRef]
53. Fayaz, G.; Goli, S.A.H.; Kadivar, M.; Valoppi, F.; Barba, L.; Calligaris, S.; Nicoli, M.C. Potential application of pomegranate seed oil oleogels based on monoglycerides, beeswax and propolis wax as partial substitutes of palm oil in functional chocolate spread. *LWT* **2017**, *86*, 523–529. [CrossRef]
54. Hwang, H.; Singh, M.; Lee, S. Properties of cookies made with natural wax–vegetable oil organogels. *J. Food Sci.* **2016**, *81*, C1045–C1054. [CrossRef] [PubMed]
55. Alvarez-Ramirez, J.; Vernon-Carter, E.J.; Carrera-Tarela, Y.; Garcia, A.; Roldan-Cruz, C. Effects of candelilla wax/canola oil oleogel on the rheology, texture, thermal properties and in vitro starch digestibility of wheat sponge cake bread. *LWT* **2020**, *130*, 109701. [CrossRef]
56. Atudorei, D.; Atudorei, O.; Codină, G.G. The Impact of germinated chickpea flour addition on dough rheology and bread quality. *Plants* **2022**, *11*, 1225. [CrossRef]
57. *ICC Standard 173*; Determination of Rheological Behaviour as a Fuction of Mixing and Temperature Increase. ICC Standards: Vienna, Austria, 2010.

gels

Article

Preparation and Characterization of Boza Enriched with Nonfat Dry Milk and Its Impact on the Fermentation Process

Ezgi Pulatsu [1,*], Sargun Malik [2], Mengshi Lin [3], Kiruba Krishnaswamy [3,4] and Bongkosh Vardhanabhuti [3]

1 School of Nutrition Sciences, Faculty of Health Sciences, University of Ottawa, Ottawa, ON K1H 8M5, Canada
2 RNAissance AG LLC, Biotechnology Research Overland Park, St. Louis, MO 63132, USA; sargunmalik@outlook.com
3 Department of Food, Nutrition & Exercise Sciences, University of Missouri, Columbia, MO 65201, USA; linme@missouri.edu (M.L.); krishnaswamyk@missouri.edu (K.K.); vardhanabhutib@missouri.edu (B.V.)
4 Department of Biomedical, Biological and Chemical Engineering, University of Missouri, Columbia, MO 65211, USA
* Correspondence: epulatsu@uottawa.ca; Tel.: +1-(613)-562-5700

Abstract: Boza is an indigenous, traditional, low-alcohol and highly viscous beverage prepared by fermenting cereals. Its thick and gel-like consistency make it suitable for consumption via spoon. Although boza is a nutritious beverage, its protein content is very low (<2%). A new type of boza was developed by incorporating nonfat dry milk (NFDM) to elevate the protein content of the beverage. Different NFDM amounts (10 to 40% w/v) were added to determine the best concentration and fermentation time based on the refractive index and pH values at room temperature (0–48 h). The best sample was further characterized by rheological analyses and Fourier transform infrared (FTIR) spectroscopy. The sample with 10% NFDM was the best, as fermentation was successfully performed, and further addition of NFDM increased the initial pH. The refractive index and pH decreased from 21.9 $\pm$ 0.1 to 11.8 $\pm$ 0.1 and 5.77$\pm$ 0.50 to 4.09 $\pm$ 0.35 during fermentation, respectively. The samples exhibited shear-thinning, solid-like behavior, and a gel-like structure. FTIR analysis by independent modeling of class analogy (SIMCA) demonstrated that unfermented slurry and the fermented product could be effectively differentiated. With the addition of 10% NFDM, the increase in the protein content of the boza medium became significant.

Keywords: colloids; fermentation; FTIR; gel; plant-based; rheology

1. Introduction

Boza is a highly viscous fermented cereal beverage that originated in Central Anatolia. It is a healthy and nutritious beverage due to its lactic acid content [1]. It also has the ability to improve digestion and intestinal flora [1]. This indigenous and traditional beverage is made from different raw materials, depending on the region or country, and it is considered as a local product [2]. Boza has been associated with many countries, including Albania, Bulgaria, Bosnia and Herzegovina, Kazakhstan, Kyrgyzstan, Macedonia, Montenegro, Romania, Serbia, and Turkey [3]. Similar millet-based fermented beverages also exist in different countries [1]. Even though modern production relies on controlled fermentation with starter cultures, most of the time such traditional foods are the products of uncontrolled and spontaneous fermentation, as has been practiced for centuries.

Boza is described as a colloidal suspension from the food science perspective. Its color ranges from light to dark beige. Its taste ranges from sweet to slightly sharp to slightly sour. It is made from wheat, rye, millet, maize, and other cereals, mixed with sugar or saccharine [4]. These types of fermented beverages can be manufactured through a set of processes such as steeping, milling, slurrying, sieving, fermentation, sedimentation, and cooking [1]. Fundamentally, the cereal is cooked for several hours and then strained to remove the solids. Then, sugar is introduced, together with a starter culture, to initiate the fermentation at 30 °C for 24 h. After the subsequent cooling process, the product can be stored for 3–5 days for consumption [5]. Owing to its lactic acid, fat, protein, carbohydrate, fiber, and vitamin contents, boza can be recognized as a valuable fermented food contributing to human nutrition. Furthermore, its processing results in improved taste and quality, together with digestible compounds, as mentioned by [6]. Boza contains approximately 0.50–1.61% protein, 12.3% carbohydrate, and 75–85% moisture. The pH range of boza samples ranges from 3.16 and 4.02, and the average alcohol content of Turkish boza samples was found to range from 0.03 to 0.39% (w/v) [7].

From a microbiological point of view, *Lactobacillus*, *Lactococcus*, *Leuconostoc*, *Pediococcus*, *Enterococcus*, *Oenococcus*, and *Weissella* are the dominant genera isolated from boza samples known to be rich in lactic acid bacteria (LAB). On the other hand, some molds and yeasts have been isolated from traditionally fermented boza, where the previous batch was used as a starter culture [8]. *Candida* and *Saccharomyces* are the two major isolated yeast genera from boza. Bulgarian boza has been studied for microflora identification in which yeasts and LAB have been identified in an average LAB/yeast ratio of 2.4 [8,9].

The pourable and semi-solid nature of boza can be characterized by rheological studies, including steady-shear and oscillatory measurements. As stated by Owens et al. [10], the study of the flow behavior of food ingredients and products has potential widespread impact on the economies of production, public health, and individual diets. Rheological study is particularly insightful for understanding and adjusting properties such as the texture, drinkability, and mouthfeel of food products with multiple ingredients and time-dependent properties [10]. The thick and gel-like consistency of this drink, with distinct characteristics, has been identified in several studies using steady-shear and oscillatory measurements [11–14]. Viscosity, storage modulus, and loss modulus are the most-used rheological parameters in these studies. Model-fitting parameters such as the consistency coefficient and the flow behavior index are also discussed. For instance, random boza samples from local markets in Turkey exhibited gel-like behavior, together with shear-thinning behavior, as described by [11].

Plant-based beverages are generally considered functional and sustainable carriers of bioactive compounds with potential health benefits [15]. However, compared to dairy products, they have poor emulsion conditions, a light and bitter taste, and insufficient nutritional components [15]. One way to alter their properties is through fermentation, in which direct metabolites and secondary metabolites are produced under the specific survival conditions of microorganisms that are beneficial to the taste, flavor, and functional properties of the beverages [15].

The increasing demand for dairy and non-dairy fermented beverages, along with the potential health implications, led to increased research and publications in this area, as pointed out by Marrero et al. [16]. However, except for the studies of Arslan-Tontul and Erbas [17] and Celik et al. [13], no other study exists about the protein enrichment of boza, even though the protein content has been reported to be low. The addition of gluten, zein, and chickpea flour improved the protein content, while volatile compounds formed due to the activity of *Lactobacillus acidophilus*, *Bifidobacterium bifidum*, and *Saccharomyces boulardii* during the co-culture fermentation of the cereal medium [17]. Owing to the addition of gluten, the protein content of the boza medium increased by up to 271% [17]. Celik et al. [13] investigated the substitution of cereal flour (maize, rice, and wheat) with yellow chickpea flour, which yielded the highest protein content (8.46%) among the samples.

Proteins are among the most important components of diet. They are a good source of energy and overall nutrition that help in maximizing the performance of the human body [18]. Since one-third of the world's population suffers from protein deficiency [17], there is a significant need to fortify food products with diverse protein sources, including plant-based and animal proteins. Animal proteins are considered the highest quality proteins with an adequate amount of essential amino acid [19]. Along with the necessary nutrition, they are also known for their functional benefits, including solubility and heat stability, as well as for their foaming and gelling properties [20]. One of the popular animal-sourced proteins is nonfat dry milk (NFDM). NFDM has a mild taste due to the lack of fat and, therefore, it is generally used to enhance the nutritional profile of non-dairy products. Therefore, the use of NFDM in boza is viable to boost the protein content of the traditional product.

Owing to their ready-to-eat forms, fermented ethnic foods and industrial probiotic foods fulfill the current requirement of delivery vehicles of bioactive compounds and nutrients at convenience [3,21,22]. Behera and Panda [21] point out that improvement of the nutritional value of ethnic foods will be the priority area in the near future. Hence, the objective of this study is to create a beverage inspired by a traditional recipe by incorporating a dairy ingredient, NFDM, that will help improve its nutritional profile, especially its protein content. The research focuses on studying the quality attributes of nonfat dry milk with boza by investigating pH and the refractive index as well as rheological and Fourier transform infrared (FTIR) spectroscopy characteristics.

2. Results and Discussion

2.1. The pH and Refractive Index Results

The change in the initial pH values of all samples is given in Figure 1a. The addition of NFDM increased the initial pH values significantly ($p < 0.05$); however, further addition of NFDM did not affect the pH values significantly ($p > 0.05$). In other words, samples with 10 to 40% NFDM had similar pH values ($p > 0.05$) (Figure 1a). For boza and other fermented beverages, measuring pH helps determine the fermentation duration and estimate the growth of LAB. In this study, pH values were used to designate the initial pH values and compare the samples with varying NFDM content.

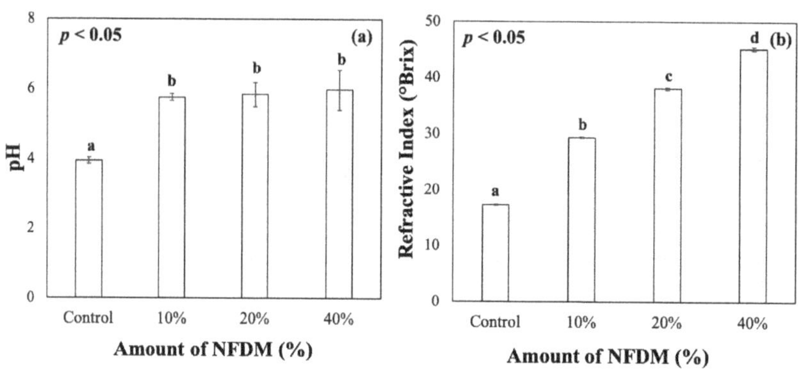

Figure 1. The effect of NFDM content on the initial pH (**a**) and RI values (**b**).

In addition to initial pH, initial RI values of the samples were determined, as shown in Figure 1b. It was found that initial brix values of the 10%, 20% and 40% (*w*/*v*) samples were 29.4, 38.1 and 45.3 as compared to 17.4 of the control sample (Figure 1b). The increase in RI was in line with the findings that the RI increases with an increase in total solids in the formulation [23].

The initial pH values were 5.77 for 10% (*w*/*v*), 5.84 for 20% (*w*/*v*) NFDM, and 5.98 for 40% (*w*/*v*) NFDM as compared to the control sample which was at 3.95. Even though the initial pH is not significantly different among the NFDM-containing samples, as given in Figure 1a, it becomes harder to accommodate the pH drop with the same amount of starter culture within the duration of fermentation. This might be because of the increased viscosity of the samples (results were not shown) when more NFDM was incorporated, which affected the microbial growth and their diffusion within the sample. The NFDM addition at 20–40% yielded very thick slurries (visual observation), which were significantly different from the consistency of the control sample (results are not shown). Therefore, based on the pH, RI values and appearance of the sample in terms of foaming and bubble formation, 10% (*w*/*v*) was selected for the formulation of the beverage. Moreover, NFDM or protein enriched beverages in the literature, is generally at 10% (*w*/*w* or *w*/*v*), which is in line with the present study [17,24].

In general, the pH of NFDM is in the range of 6.5–6.7 and the titratable acidity varies from 0.1 to 0.15%. In a study conducted by [25], the relationship between titratable acidity and pH was observed in a fermentation process of nonfat milk. It was observed that during the fermentation, the acidity increases for every component of the powder that is reconstituted [25,26]. Lactic acid bacteria grow readily in the food substrates and lower the pH rapidly to a point where competing organisms are no longer able to grow. They help in the preservation with comparatively lower energy requirements and consequently lower cost. The preservation occurs with nicotinamide adenine dinucleotide (NADH) oxidation that further reacts with oxygen [27]. During fermentation, free fatty acids are formed due to the lipolysis of fat in the bulgur grain, which contributes to decreased pH. The results were similar to a study conducted wherein cereal grains such as sorghum, maize and other millets were fermented using LAB and pH was found to be a critical indicator for estimating the LAB growth [28].

During fermentation, both traditional boza and dairy-based boza beverages exhibited a decrease in pH, with the traditional beverage sample showing a rapid drop compared to dairy-based sample. After 24 h, the pH values were 4.49 and 6.28 for traditional and dairy-based boza. There was no significant change in pH for the dairy-based sample after 24 h. After 48 h, the dairy-based sample went through a similar pH drop wherein both the samples had a pH value of 4.09, as shown in Figure 2a, indicating a slow fermentation in the dairy-based sample. The RI values decreased in both traditional boza and dairy-based boza during fermentation Figure 2b. The initial RI for traditional and dairy-based samples were noted as 12.7 and 21.9° Brix, respectively. After 24 h fermentation, a drop in RI was observed in the traditional boza sample, but that of the dairy-based sample remained similar to its initial value. As depicted in Figure 2b, after 48 h fermentation, a drastic change was observed in the dairy-based sample, wherein it reached 11.8° Brix, which was similar to the traditional sample. The RI values obtained in the experiment were indicative of the composition and soluble dry matter content of the sample. During fermentation, microorganisms use sugars and convert them to ethyl alcohol and carbon dioxide. Hence, the RI could be used to represent the progress of the fermentation process. A similar method was used for analyzing the fermentation process in a rice-based fermented product called *Tape Ketan* in a study conducted by [29]. RI was used for alcohol measurement, and the study concluded that as the fermentation progressed, there was an increase in alcohol content and a decrease in dissolved sugar, justifying the reduction in RI.

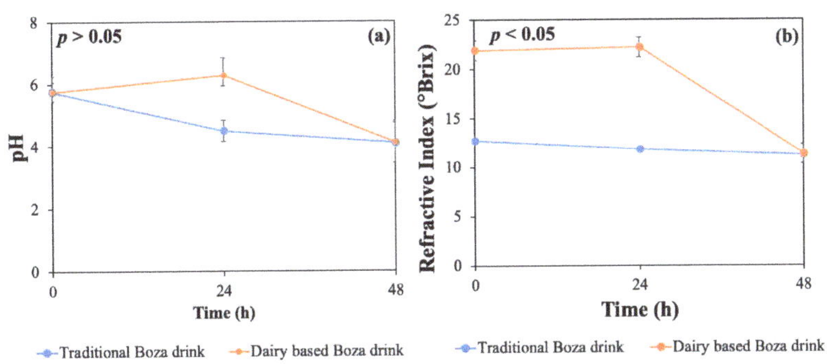

Figure 2. Change in pH values of traditional and 10% (*w*/*v*) NFDM dairy-based boza during fermentation (**a**); change in RI values of traditional and 10% (*w*/*v*) NFDM dairy-based boza during fermentation (**b**).

RI also suggested the amount of NFDM that could be added in the traditional beverage to obtain a dairy-based beverage. Hence, the sample containing 10% (w/v) NFDM was finalized for further experimentation. Given that NFDM contains 34% protein, the protein content could be increased significantly.

2.2. Rheological Properties of Dairy-Based Slurry and Boza

As shown in Figure 3a, the apparent viscosity of fermented (SDB-F) boza and unfermented dairy-based slurry (SDB) showed shear-thinning behavior evident by the reduction in the apparent viscosity values over the increased shear rate. The related model parameters were tabulated in Table 1. The correlation coefficients for the model fitting were greater than 0.92, which ensures a successful data fitting. Both samples exhibited similar behavior in the given range of shear rates ($p \leq 0.05$). Fermentation did not affect the flow behavior of the sample significantly ($p \leq 0.05$). Also, n values were less than 1, which proved the shear-thinning behavior of the samples (Table 1). Hence, both consistency and the flow behavior index were not affected by fermentation, which might be related to low alcohol content and CO_2 levels reached during fermentation.

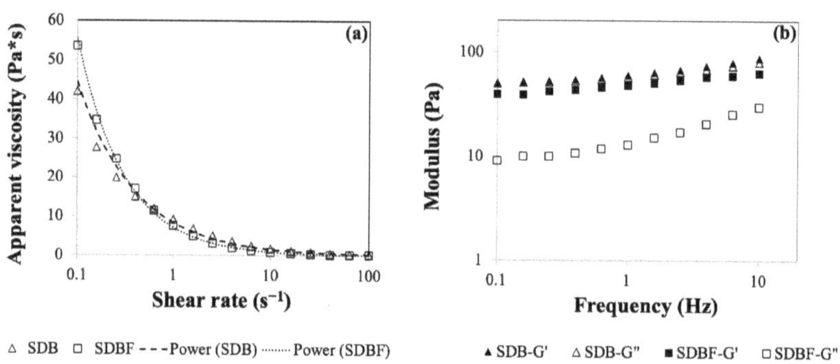

Figure 3. Apparent viscosity vs. shear rate (**a**) and frequency sweep results (**b**) of dairy-based boza and unfermented slurry.

Table 1. The results of model fitting for viscometry and oscillation. (Means sharing the same letters (a) are statistically the same ($p \geq 0.05$). Each parameter was analyzed separately).

Sample	Model Fitting Parameters for Viscometry and Oscillation		
	Model fitting parameters for viscometry		
	K	n	R^2
SDB-F	7.10 ± 0.43 [a]	0.11 ± 0.05 [a]	0.92
SDB	8.36 ± 0.38 [a]	0.28 ± 0.01 [a]	0.98
	Model fitting parameters for oscillation		
	a′	b′	R^2
SDB-F	41.16 ± 17.72 [a]	1.39 ± 0.17 [a]	0.96
SDB	47.04 ± 17.52 [a]	1.35 ± 0.04 [a]	0.93
	Model fitting parameters for oscillation		
	a″	b″	R^2
SDB-F	39.97 ± 17.57 [a]	1.35 ± 0.16 [a]	0.94
SDB	47.04 ± 17.52 [a]	1.30 ± 0.03 [a]	0.93

In the literature, the rheological properties of boza have been studied, and similar findings have been reported [12]. However, there is no report on the characterization of the slurry before fermentation. The current study indicated no significant change in the flow properties of the slurry and fermented product.

After characterization of the viscosity of the samples, dynamic rheological testing was performed, as shown in Figure 3b. The storage and loss modulus of the samples were plotted against the frequency range. The results demonstrated that for the given frequency range, solid-like behavior dominated the overall behavior since $G' > G''$, for both samples (Figure 3b). It can be inferred that the samples are gel-like structures with limited frequency dependency in the given frequency range. The viscoelastic nature of the samples was successfully captured via dynamic rheological measurements aligning with the literature reports [11]. Additionally, model fitting enabled sample comparison to designate if there is a significant difference using statistical analysis. No significant difference was found between the samples at the 0.05 significance level, which aligned with the viscometry results. Both fermented product and unfermented slurry exhibited similar rheological behavior.

2.3. FTIR Spectroscopy

Figure 4 presents FTIR spectra and related SIMCA test results to differentiate the fermented sample (SDB-F) from the unfermented sample (SDB). Two major peaks, 1638 and 3302 cm^{-1}, observed within the scanned range (Figure 4a). The first peak can be correlated with the amide bands corresponding to proteins, or more specifically C=O stretching vibrations of amides of proteins [30]. Kher et al. [31] reported that the bands in the amide region (1700–1600 cm^{-1}) have previously been assigned to the α-helix and that at approximately 1630 cm^{-1} to the β-sheet by several authors [32]. The 1638 cm^{-1} can be attributable to β-sheet structures resulting from NFDM. Hence, the findings were in good agreement with the literature [31]. The latter peak can be associated with O–H stretch of water.

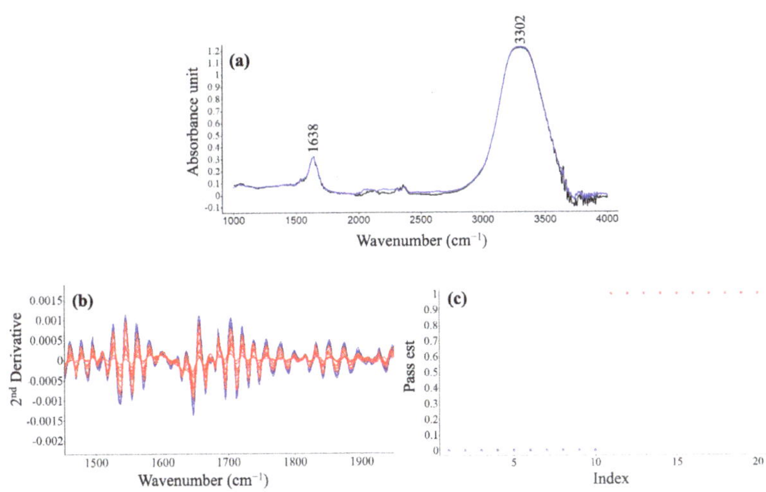

Figure 4. FTIR absorbance spectra of SDB-F (blue) and SDB (black) representing the average of ten sample measurements (**a**), the second derivative of all spectra: red represents SDB, and blue represents SDB-F (**b**), and SIMCA result: blue (10 replicates of SDB-F sample spectra) and blue (10 replicates of SDB sample spectra) (**c**).

Several alcoholic compounds can be formed during cereal fermentation, n-propanol, isobutanol, amyl alcohol, isoamyl alcohol, 2,3-butandieol, and β-phenylethyl alcohol [33]. Three specific spectral regions of 3005–2960 cm^{-1}, 1200–950 cm^{-1} and 900–850 cm^{-1} were reported to be ethanol-related peaks [34].

The second derivative spectra were analyzed to elucidate small spectral differences among the samples, as shown in Figure 4b. The intensities from the fermented product were stronger than the unfermented slurry in the scanning range (Figure 4b). Furthermore, SIMCA was applied to the spectra using a pass test, where the pass value will be true (1) or false (0) according to the base spectra selected. The results indicated that SDB-F and SDB samples are different from each other, as expected, due to several biochemical changes due to fermentation. All ten replicates of each group were successfully discriminated against via SIMCA.

This study demonstrates that FTIR spectroscopy, in conjunction with SIMCA, can consistently differentiate samples of fermented dairy-based boza and unfermented dairy-based slurry. However, a more detailed analysis should be performed to designate biochemical changes during fermentation which will be performed in future studies.

3. Conclusions

The incorporation of NFDM was successfully achieved to obtain a dairy-based boza, leading to increase in protein content given that NFDM contains 34% protein. However, the addition of NFDM resulted in an increased initial pH of the slurry that prolonged the fermentation time to 48 h. Both pH and RI values decreased during the course of fermentation, as expected. Based on the physicochemical analyses, the sample with 10% NFDM was designated as the best sample. According to the results of rheological characterization, the samples are shear-thinning, solid-like behavior, and gel-like structures. FTIR analysis, together with chemometrics methods, can differentiate the unfermented slurry from the fermented product up to 100% of the time. Thus, this method has great potential to study cereal-based fermented beverages.

This study is a step toward the research efforts in the development of enriched food products with diverse ingredients. Millet-based agricultural products are abundant, and their utilization in various food products is nutritionally and economically promising. New consumer trends are in the direction of ethnic/traditional foods and beverages; therefore, dairy-based boza can be a good alternative for finding healthy, natural, and nutritious ethno-food. This preliminary study is unique in terms of evaluating the suitability of NFDM addition into plant-based boza beverage. Further research is necessary to analyze the biochemistry profile as well as the protein digestibility of this unique fermented dairy-based beverage. Also, more characterization techniques should be employed to discuss the properties of the product further, including sensory properties.

4. Materials and Methods

4.1. Materials

Sugar and nonfat dry milk (instant dry milk, Great Value) were purchased from the local market (Columbia, MO, USA). Bulgur (Duru, fine bulgur, 11 g protein/100 g) and a commercial boza were supplied from Turkey (Omur Boza).

4.2. Boza Preparation

With an extensive search in the literature and preliminary analyses, the boza recipe was established. Briefly, 1 cup of bulgur (160 $\pm$ 10 g) was soaked in water overnight. Then, the soaked bulgur was boiled in 1.5 L of distilled water for 1.5 h. Half of the bulgur grains were removed to obtain better consistency, and the remaining mixture was passed through the sieve (1 mm), which was designated by preliminary experiments. The obtained solution was blended with a kitchen hand blender for 1 min and cooled down to 40 °C in the refrigerator. To the blended mixture, 20 g of boza as a starter culture, 37.5 g of sugar and 100 mL of distilled water were added. The whole mixture was then left for fermentation at room temperature for 48 h.

4.3. Dairy-Based Boza Preparation

Similar to traditional boza preparation, dairy-based boza was obtained, as shown in Figure 5. Additionally, 50 g of NFDM was mixed with 100 mL water to obtain the dairy-based slurry. The mixture was then boiled for 10–15 min and cooled down to 40 °C. Sugar (37.5 g) was then added together with 20 g of boza starter culture. The mixture was then left for fermentation at room temperature for 48 h. Upon completion of fermentation, samples were placed in the refrigerator for 24 h prior (which is a common practice before the consumption of the fermented beverage) to rheological and spectral analyses.

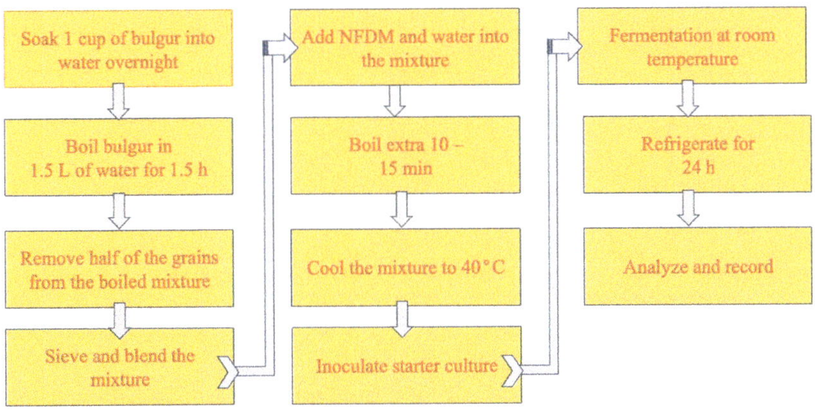

Figure 5. Flow diagram of dairy-based boza production.

4.4. The pH and Refractive Index Measurements

The pH of the samples was measured using a digital pH meter (Mettler Toledo™, Columbus, OH, USA) with a pH electrode (In Lab® Expert Pro-ISM, Columbus, OH, USA). The refractive index (RI) was measured using digital refractometer (WM-7 ATAGO™, Bellevue, WA, USA), which provides information regarding the water-soluble dry matter content. Traditional boza was used as a reference sample to determine the fermentation time for dairy-based boza, as this is the first study exploring dairy-based boza. The pH and RI of traditional and dairy-based boza were measured at 0, 24 and 48 h after fermentation. Then, the fermentation time was designated when the Brix° and pH were the same for both traditional and dairy-based boza samples.

4.5. Rheological Measurements

Rheological characterization of selected samples was carried out using a Kinexus Pro rheometer (NETZSCH Instruments North America, LLC, Burlington, MA, USA) operated with a parallel plate geometry of 50 mm diameter and a 2 mm gap. Fermented dairy-based boza (FDB) and unfermented dairy-based slurry (DB) (before the addition of starter culture) were selected for the analysis after successful trials. The samples were taken out to room temperature (25 ± 1 °C) for one hour before the rheological analysis. Viscosity measurements were carried out in the shear rate range of 0.1–100 s^{-1}. All measurements were performed in triplicate. The power-law model fitting was carried out to describe the flow behavior of the samples (Equation (1)).

$$\sigma = \eta \dot{\gamma}^{n} \quad (1)$$

where σ represents stress (Pa), η is the viscosity, $\dot{\gamma}$ represents the shear rate, and n is the power of the rate. Then, the equation was converted to Equation (2) to represent the K

(Pa $*$ s^n) and n (dimensionless) parameters which are the consistency coefficient and the flow behavior index, respectively [35].

$$\eta = (K\dot{\gamma})^{n-1} \tag{2}$$

For the dynamic rheological properties of the products, amplitude sweep tests (at 1 Hz in the range of 0.01–100% strain) were run to determine the linear viscoelastic region (LVER) prior to frequency sweep measurements. Frequency tests were performed at 0.1% shear strain in the frequency range of 1–10 Hz, where storage (G') and loss (G'') modulus were recorded to describe the rheological behavior of the samples.

G' and G'' values obtained from the frequency tests were fitted in the model as described in the following equations.

$$\log G' = \log a' + b' \log \omega \tag{3}$$

$$\log G'' = \log a'' + b'' \log \omega \tag{4}$$

where ω rad/s is the angular frequency and a', a'', b', and b'' are the model parameters. The frequency dependence of the samples was quantified via these model parameters. Among them, a' and a'' are the flow coefficients, b' and b'' indicate the frequency sensitivity of the samples [36].

4.6. FTIR Analysis of Dairy-Based Boza

FTIR spectra of selected samples, SDB-F and SDB, were collected by an FTIR spectrometer (Nicolet 380; Thermo Fischer Scientific, Waltham, MA, USA), in the region of (4000–400 cm^{-1}) with attenuated total reflection mode. FTIR spectra (n = 10) of each sample were collected where the resolution was set as 2 cm^{-1} with an average of 32 scans for each spectrum to be analyzed using Delight version 3.2.1 (D-Squared Development, LaGrande, OR, USA) software. First, data processing was performed, where spectral data were merged and smoothed with a Gaussian function (over 12 cm^{-1}). Then, a second derivative transformation with a 12 cm^{-1} gap was performed [30]. SIMCA classification was performed on all the spectral data collected.

4.7. Data Analysis and Statistics

Experimental data were given as the mean $\pm$ standard deviation for pH, RI and rheological parameters. The significant differences among the data points were evaluated based on one way analysis of variance (ANOVA) and Tukey's test (Minitab, Version 16.2.0.0, Minitab Inc., Coventry, UK) with a p-value of $\leq$0.05. For the spectral analysis, SIMCA was used, which is a classification method based on principal component analysis (PCA), a common multivariate statistical technique.

Author Contributions: E.P., conceptualization, methodology, data curation, formal analysis, writing—original draft. S.M., Methodology, data curation, formal analysis, writing—original draft. B.V., supervision, resources, writing—review and editing. M.L., formal analysis, resources, writing—review and editing. K.K., resources, writing—review and editing. All authors have read and agreed to the published version of the manuscript.

Funding: This research was partially supported by the Robert T. Marshall Scholarship for the first author as well as the USDA NIFA 2020-67030-31336.

Data Availability Statement: The data presented in this study is already available in the manuscript. Additional information can be provided by the corresponding author upon request.

Conflicts of Interest: The study was done when Sargun Malik was a PhD student at University of Missouri, who is currently working for a company. There is no relationship with the company and the products, materials, etc.

References

1. Amadou, I. Millet Based Fermented Beverages Processing. In *Fermented Beverages*; Elsevier: Amsterdam, The Netherlands, 2019; pp. 433–472.
2. Setyowati, N.; Masyhuri; Mulyo, J.H.; Irham; Yudhistira, B. The Hidden Treasure of Wedang Uwuh, an Ethnic Traditional Drink from Java, Indonesia: Its Benefits and Innovations. *Int. J. Gastron. Food Sci.* **2023**, *31*, 100688. [CrossRef]
3. Petrova, P.; Petrov, K. Traditional Cereal Beverage Boza: Fermentation Technology, Microbial Content and Healthy Effect. In *Fermented Foods Part II Technological Interventions*; Ray, R.C., Montet, D., Eds.; CRC Press Taylor & Franchis Group, LCC: Boca Raton, FL, USA, 2017; pp. 284–305.
4. Vaudha, S.; Mishra, H. Non Dairy Probiotic Beverages. *Int. Food Res. J.* **2013**, *20*, 7–15.
5. Hancioğlu, Ö.; Karapinar, M. Microflora of Boza, a Traditional Fermented Turkish Beverage. *Int. J. Food Microbiol.* **1997**, *35*, 271–274. [CrossRef] [PubMed]
6. Leblanc, J.G.J.; Todorov, S.D. Bacteriocins Producing Lactic Acid Bacteria Isolated from Boza, a Traditional Fermented Beverage from Balkan Peninsula—From Isolation to Application. In *Science against Microbial Pathogens: Communicating Current Research and Technological Advances*; Formatex Research Center: Badajoz, Spain, 2011; pp. 1311–1320. ISBN 978-84-939843-2-8.
7. Yeğin, S.; Üren, A. Biogenic Amine Content of Boza: A Traditional Cereal-Based, Fermented Turkish Beverage. *Food Chem.* **2008**, *111*, 983–987. [CrossRef]
8. Caputo, L.; Quintieri, L.; Baruzzi, F.; Borcakli, M.; Morea, M. Molecular and Phenotypic Characterization of Pichia Fermentans Strains Found among Boza Yeasts. *Food Res. Int.* **2012**, *48*, 755–762. [CrossRef]
9. Blandino, A.; Al-Aseeri, M.E.; Pandiella, S.S.; Cantero, D.; Webb, C. Cereal-Based Fermented Foods and Beverages. *Food Res. Int.* **2003**, *36*, 527–543. [CrossRef]
10. Owens, C.E.; Fan, M.R.; Hart, A.J.; McKinley, G.H. On Oreology, the Fracture and Flow of "Milk's Favorite Cookie®". *Phys. Fluids* **2022**, *34*, 043107. [CrossRef]
11. Muhammet, A.; Ersoz Tatlisu, N.B.; Toker, Ö.S.; Yilmaz, M.T.; Cankurt, H.; Durak, M.Z.; Sagdic, O. Microbiological, Steady, and Dynamic Rheological Characterization of Boza Samples: Temperature Sweep Tests and Applicability of the Cox–Merz Rule. *Turk. J. Agric. For.* **2014**, *38*, 377–387. [CrossRef]
12. Genç, M.; Zorba, M.; Ova, G. Determination of Rheological Properties of Boza by Using Physical and Sensory Analysis. *J. Food Eng.* **2002**, *52*, 95–98. [CrossRef]
13. Çelik, İ.; Işik, F.; Yilmaz, Y. Effect of Roasted Yellow Chickpea (Leblebi) Flour Addition on Chemical, Rheological and Sensory Properties of Boza. *J. Food Process. Preserv.* **2016**, *40*, 1400–1406. [CrossRef]
14. Göncü, A.; Kuzumoğlu, Y.; Vardin, A.Y.; Çelik, İ.; Karaman, A.D. Evaluation of Physicochemical and Sensorial Properties of Gluten-free Boza Formulated with Chia and Quinoa Flours. *Cereal Chem.* **2023**, *100*, 1123–1131. [CrossRef]
15. Liu, H.; Xu, X.; Cui, H.; Xu, J.; Yuan, Z.; Liu, J.; Li, C.; Li, J.; Zhu, D. Plant-Based Fermented Beverages and Key Emerging Processing Technologies. *Food Rev. Int.* **2023**, *39*, 5844–5863. [CrossRef]
16. Marrero, S.C.; Martínez-Rodríguez, A.; Pérez, S.E.M.; Moya, S.P. New Trends and Applications in Fermented Beverages. In *Fermented Beverages*; Elsevier: Amsterdam, The Netherlands, 2019; pp. 31–66.
17. Arslan-Tontul, S.; Erbas, M. Co-Culture Probiotic Fermentation of Protein-Enriched Cereal Medium (Boza). *J. Am. Coll. Nutr.* **2020**, *39*, 72–81. [CrossRef] [PubMed]
18. Weijs, P.J.; Cynober, L.; DeLegge, M.; Kreymann, G.; Wernerman, J.; Wolfe, R.R. Proteins and Amino Acids Are Fundamental to Optimal Nutrition Support in Critically Ill Patients. *Crit. Care* **2014**, *18*, 591. [CrossRef] [PubMed]
19. Hackney, K.J.; Trautman, K.; Johnson, N.; Mcgrath, R.; Stastny, S. Protein and Muscle Health during Aging: Benefits and Concerns Related to Animal-Based Protein. *Anim. Front.* **2019**, *9*, 12–17. [CrossRef] [PubMed]
20. Wilding, P.; Lillford, P.J.; Regenstein, J.M. Functional Properties of Proteins in Foods. *J. Chem. Technol. Biotechnol. Biotechnol.* **1984**, *34*, 182–189. [CrossRef]
21. Behera, S.S.; Panda, S.K. Ethnic and Industrial Probiotic Foods and Beverages: Efficacy and Acceptance. *Curr. Opin. Food Sci.* **2020**, *32*, 29–36. [CrossRef]
22. Ferreira, I.; de Sousa Melo, D.; Menezes, A.G.T.; Fonseca, H.C.; de Assis, B.B.T.; Ramos, C.L.; Magnani, M.; Dias, D.R.; Schwan, R.F. Evaluation of Potentially Probiotic Yeasts and Lactiplantibacillus Plantarum in Co-Culture for the Elaboration of a Functional Plant-Based Fermented Beverage. *Food Res. Int.* **2022**, *160*, 111697. [CrossRef]
23. Meichner, C.; Schedl, A.E.; Neuber, C.; Kreger, K.; Schmidt, H.-W.; Kador, L. Refractive-Index Determination of Solids from First- and Second-Order Critical Diffraction Angles of Periodic Surface Patterns. *AIP Adv.* **2015**, *5*, 087135. [CrossRef]
24. Choi, I.; Li, N.; Vuia-Riser, J.; Carter, B.; Drake, M.; Zhong, Q. Neutral PH Nonfat Dry Milk Beverages with Turbidity Reduced by Sodium Hexametaphosphate: Physical and Sensory Properties during Storage. *LWT* **2021**, *147*, 111656. [CrossRef]
25. Wilkowske, H.H. Relationship between Titratable Acidity and PH during Lactic Acid Fermentation in Reconstituted Nonfat Milk. *J. Dairy Sci.* **1954**, *37*, 22–29. [CrossRef]
26. FAO. *Off-Tastes in Raw and Reconstituted Milk*; FAO: Roma, Italy, 2021; ISBN 92-5-101258-X.
27. Steinkraus, K.H. Lactic Acid Fermentations. In *Applications of Biotechnology to Traditional Fermented Foods*; National Academy Press: Washington, DC, USA, 1992; ISBN 0309046858.
28. Nwachukwu, E.; Achi, O.; Ijeoma, I. Lactic Acid Bacteria in Fermentation of Cereals for the Production of Indigenous Nigerian Foods. *Afr. J. Food Sci. Technol.* **2010**, *1*, 21–26.

29. Muchtaridi, M.; Musfiroh, I.; Hambali, N.N.; Indrayati, W. Determination of Alcohol Contents of Fermentated Black Tape Ketan Based on Different Fermentation Time Using Specific Gravity, Refractive Index and GC-MS Methods. *J. Microbiol. Biotechnol. food Sci.* **2012**, *2*, 933–946.
30. Lin, M.; Al-Holy, M.; Chang, S.-S.; Huang, Y.; Cavinato, A.G.; Kang, D.-H.; Rasco, B.A. Rapid Discrimination of Alicyclobacillus Strains in Apple Juice by Fourier Transform Infrared Spectroscopy. *Int. J. Food Microbiol.* **2005**, *105*, 369–376. [CrossRef] [PubMed]
31. Kher, A.; Udabage, P.; McKinnon, I.; McNaughton, D.; Augustin, M.A. FTIR Investigation of Spray-Dried Milk Protein Concentrate Powders. *Vib. Spectrosc.* **2007**, *44*, 375–381. [CrossRef]
32. Markoska, T.; Huppertz, T.; Grewal, M.K.; Vasiljevic, T. FTIR Analysis of Physiochemical Changes in Raw Skim Milk upon Concentration. *LWT* **2019**, *102*, 64–70. [CrossRef]
33. Kohajdova, Z.; Karocicova, J. Fermentation of Cereals for Specific Purpose. *J. Food Nutr. Res.* **2007**, *46*, 51–57.
34. Debebe, A.; Redi-Abshiro, M.; Chandravanshi, B.S. Non-Destructive Determination of Ethanol Levels in Fermented Alcoholic Beverages Using Fourier Transform Mid-Infrared Spectroscopy. *Chem. Cent. J.* **2017**, *11*, 27. [CrossRef]
35. Tekin Pulatsü, E.; Sahin, S.; Sumnu, G. Characterization of Different Double-Emulsion Formulations Based on Food-Grade Emulsifiers and Stabilizers. *J. Dispers. Sci. Technol.* **2018**, *39*, 996–1002. [CrossRef]
36. Pulatsu, E.; Su, J.W.; Lin, J.; Lin, M. Factors Affecting 3D Printing and Post-Processing Capacity of Cookie Dough. *Innov. Food Sci. Emerg. Technol.* **2020**, *61*, 102316. [CrossRef]

MDPI AG
Grosspeteranlage 5
4052 Basel
Switzerland
Tel.: +41 61 683 77 34

Gels Editorial Office
E-mail: gels@mdpi.com
www.mdpi.com/journal/gels

www.ingramcontent.com/pod-product-compliance
Lightning Source LLC
Chambersburg PA
CBHW062050220926
51850CB00010B/2715
9783725859375